Automation, Collaboration, & E-Services

Volume 3

Series editor

Shimon Y. Nof, Purdue University, West Lafayette, USA
e-mail: nof@purdue.edu

About this Series

The Automation, Collaboration, & E-Services series (ACES) publishes new developments and advances in the fields of Automation, collaboration and e-services; rapidly and informally but with a high quality. It captures the scientific and engineering theories and techniques addressing challenges of the megatrends of automation, and collaboration. These trends, defining the scope of the ACES Series, are evident with wireless communication, Internetworking, multi-agent systems, sensor networks, cyber-physical collaborative systems, interactive-collaborative devices, and social robotics etc. Within the scope of the series are monographs, lecture notes, selected contributions from specialized conferences and workshops.

More information about this series at http://www.springer.com/series/8393

Mohsen Moghaddam · Shimon Y. Nof

Best Matching Theory & Applications

Springer

Mohsen Moghaddam
School of Industrial Engineering, PRISM Center
Purdue University
West Lafayette, IN
USA

Shimon Y. Nof
School of Industrial Engineering, PRISM Center
Purdue University
West Lafayette, IN
USA

ISSN 2193-472X ISSN 2193-4738 (electronic)
Automation, Collaboration, & E-Services
ISBN 978-3-319-46069-7 ISBN 978-3-319-46070-3 (eBook)
DOI 10.1007/978-3-319-46070-3

Library of Congress Control Number: 2016951672

Printed on acid-free paper

This Springer imprint is published by Springer Nature
The registered company is Springer International Publishing AG
The registered company address is: Gewerbestrasse 11, 6330 Cham, Switzerland

This book is dedicated to:

My parents, Eshrat and Ahmad
— Mohsen Moghaddam

My parents, Yaffa and Yaacov Nowomiast
of blessed memory
— Shimon Y. Nof

The original version of the book was revised:
For detailed information please see Erratum.
The Erratum to this book is available
at 10.1007/978-3-319-46070-3_9

Foreword

Life is complicated. I have just returned from taking breakfast in the hotel where I am staying. So many choices: Which of the three restaurants should I choose? How do I select what I want to eat from the glittering array of food? What criteria do I use to help me in my selection? What model do I use? Do I think forward to what I will have for lunch? Do I look across the table and consider what comments my wife will make of my choice? So complicated. And this is after I have had to match what I want to wear to go down to breakfast to the limited choice in my suitcase! From cradle to grave life is about choices, matching aspiration and desire to available resources according to opportunity within the context of the time.

And if making decisions is hard enough for an individual, how much harder are they for organisations, any organisations, with their distributed systems, their multiple objectives, their choice of resources, and their various constraints? How does an organisation find a "Best Match"?

This book is here to help in making such decisions. Various authors have been writing about decision-making and matching resources to desired outputs for over 60 years, but this book provides the most comprehensive source of information available to date of how the problems of matching can be classified and the wide range of techniques, approaches and modelling which can be applied to achieving a best match. The book addresses centralised, decentralised and distributed systems under static and dynamic conditions and its applications and examples range from manufacture, robotics, warehousing and supply chains through to representative systems in the various service industries. It is a book which should have wide appeal to both researchers and practitioners. Every chapter is comprehensively referenced and the practice problems presented in each chapter are well considered to reinforce the understanding and familiarity with the concepts and techniques being explored.

The analyses presented throughout the book are all based upon the PRISM taxonomy developed by the authors over a number of years at Purdue University. This provides a comprehensive reference model which can be applied to an almost universal set of circumstances. For readers not familiar with the authors' previous work it will be necessary to study carefully the description of the taxonomy given in Chap. 2. The effort is well worth the trouble as it gives the reader a powerful tool for understanding, classifying and describing complex problems and scenarios well beyond the scope of this book.

Under the umbrella of the PRISM taxonomy, the book covers a wide range of mixed-integer models and programs for formulating various instances of matching with certain dimensions, conditions, and objectives. The book poses fundamental questions on the notion of distributed decision-making—enabled by the emerging information and operational technologies—and its impact on the "quality of match" in terms of agility, fault tolerance, and stability. A state-of-the-art collection of matching tools is created by the authors, which provides step-by-step instructions on how real-time optimization and control of matching processes can be achieved. This powerful portfolio enhances learning of various exact, heuristic, and bio-inspired algorithms, agent-based concepts and methodologies, and interaction protocols, supported by several use cases.

In the final chapter of the book the authors consider the frontiers in best matching, dealing with the challenges arising from new technologies, an ever widening set of applications and developments in the tools available to tackle such problems. The chapter serves to emphasise not just the value of the book in suggesting more effective ways of addressing today's problems but its value in giving the reader confidence to address tomorrow's matching problems in a rapidly changing and uncertain world.

Chris O'Brien OBE
Emeritus Professor, Nottingham University
UK, Malaysia, China

Preface

Matching is always part of life. No one likes to be mismatched, and no one wants to be poorly matched. In playing "musical chairs," children try to catch a seat when the music stops, and there are always some that are left unmatched, without a seat. Is it just a game? Throughout life, we try to match ourselves with the right partners, the right friends, the right places, and the right programs.

Our interest in matching began in the 1990s, soon after we established the PRISM Center for Production, Robotics, and Integration Software for Manufacturing and Management, at Purdue University. Through industry projects, we identified two related problems that appeared commonly: The process of matching (the procedure, algorithm, protocol of matching), and the result of the matching (the match, sometimes called "the catch"). As industrial engineers, we sought to optimize both, looking for the best matching process, and for the best match.

Two prominent example areas that engaged us were: The selective matching of components in construction and assembly; and the matching of suppliers (providers) to clients (consumers). The purpose in both these areas is similar: To assure the best quality and reliability; resilience, and sustainability. In assembly – matching in simpler assemblies, such as hand-tools, and matching in complex assemblies, such as jet engines. Similarly, in team-building – we are looking to match the right suppliers of needed knowledge or of demanded parts to corresponding clients.

Increasingly, it became evident that there are fascinating art and science in matching. This book attempts to present a comprehensive look at what has been discovered and developed, validated, and implemented in this area, by many others and by our PRISM team, both at the PRISM Center and throughout our PGRN, the PRISM Global Research Network. There are also other types of matching and best matching that are beyond the scope of this book, such as in forensics, database search, visual and musical pattern matching, and social matching. In all these areas, best matching and best match have similar meaning, significance, and value.

We owe thanks and gratitude to our many students, colleagues, visiting scholars, and industry partners, who have participated in the development of this exciting field of study. Many of them are mentioned throughout this book. Our thanks also go to the Springer colleagues, who have helped us bring this book to light. It is part of the Springer's ACES book series, which is dedicated to Automation, Collaboration, and E-Services. As it becomes clear in reading this book, best matching is essential for automation, for collaboration, and for e-services.

We hope this book will inspire you to get as excited as we are about best matching. There are still open and fascinating questions about it, for further exploration and discovery, some of which we describe in this book.

West Lafayette
Indiana
Summer 2016

Mohsen Moghaddam
Shimon Y. Nof

About the Book

Mismatch or best match? This book demonstrates that best matching of individual entities to each other is essential to ensure smooth conduct and successful competitiveness in any distributed system, natural and artificial. Interactions must be optimized through best matching in planning and scheduling, enterprise network design, transportation and construction planning, recruitment, problem solving, selective assembly, team formation, sensor network design, and more. Fundamentals of best matching in distributed and collaborative systems are explained by providing:

- Methodical analysis of various multidimensional best matching processes
- Comprehensive taxonomy, comparing different best matching problems and processes
- Systematic identification of systems' hierarchy, nature of interactions, and distribution of decision-making and control functions
- Practical formulation of solutions based on a library of best matching algorithms and protocols, ready for direct applications and apps development

Designed for both academics and practitioners, oriented to systems engineers and applied operations researchers, diverse types of best matching processes are explained in production, manufacturing, business and service, based on a new reference model developed at Purdue University PRISM Center: "The PRISM Taxonomy of Best Matching". The book concludes with major challenges and guidelines for future basic and applied research in the area of best matching.

Contents

1 Introduction: Best Matching and Best Match 1
1.1 What Is Best Matching? 1
1.2 Definitions and Scope 4
1.2.1 Distributed Systems 6
1.2.2 *Collaboration Versus Competition* 8
1.3 Best Matching in Practice 9
1.4 Summary 12
References 14

2 The PRISM Taxonomy of Best Matching 19
2.1 Framework 19
2.1.1 D1: Sets of Individuals 21
2.1.2 D2: Matching Conditions 22
2.1.3 D3: Matching Criteria 26
2.1.4 D+: Time or Progression 29
2.1.5 The Prismatic Structure of the PRISM Taxonomy 30
2.2 Four Examples of the PRISM Taxonomy Application 30
2.2.1 Balancing Collaborative Assembly Lines ($M:1/RC,PR,RS/--,WS$) 30
2.2.2 Part Pairing for Concurrent Loading-Machining ($1:1//-,OS$) 31
2.2.3 Dynamic Teaming with Interdependent Preferences ($M:1/RC,IP/+,OS/DI,ES$) 33
2.2.4 Location-Allocation Decisions in CNO ($1:M:M/RC,PR,RS/+-,WS$) 35
2.3 Summary 37
References 41

3 Mathematical Models of Best Matching 43
3.1 Why Mathematical Modeling for Best Matching? 43
3.2 D1. Sets 45
3.2.1 One-to-One Matching 46

3.2.2 Generalized Matching . . . 47
3.2.3 Multi-Dimensional Matching . . . 48
3.3 D2. Conditions . . . 49
3.3.1 Resource-Constrained Matching . . . 49
3.3.2 Matching with Precedence Relations . . . 51
3.3.3 *Matching with Resource Sharing* . . . 51
3.3.4 Matching with Interdependent Preferences . . . 53
3.3.5 Layered Matching . . . 56
3.4 D3. Criteria . . . 57
3.5 D+. Static Versus Dynamic Matching . . . 59
3.6 Summary . . . 60
References . . . 62

4 Distributed Decision-Making and Best Matching . . . 63
4.1 Single Versus Multiple Decision-Makers . . . 63
4.2 Distribution of Decisional Abilities . . . 65
4.2.1 Example 1: Intelligent Warehouse Management Systems . . . 68
4.2.2 Example 2: Precision Agriculture . . . 70
4.2.3 Alternative Configurations—Advantages and Limitations . . . 72
4.3 Nature of Interactions . . . 74
4.4 Summary . . . 76
References . . . 79

5 Static and Centralized Matching . . . 81
5.1 Motivation for Using Algorithms . . . 81
5.2 Heuristics and Exact Algorithms . . . 83
5.2.1 Hungarian Method . . . 83
5.2.2 Deferred Acceptance Algorithm . . . 86
5.2.3 Lagrangian Relaxation Method . . . 89
5.2.4 Branch-and-Bound Method . . . 98
5.3 Metaheuristics . . . 102
5.3.1 *Genetic Algorithm (GA)* . . . 103
5.3.2 Greedy Randomized Adaptive Search Procedure (GRASP) . . . 107
5.3.3 *Ant Colony Optimization (ACO)* . . . 110
5.3.4 Tabu Search . . . 114
5.4 Summary . . . 118
References . . . 122

6 Dynamic and Distributed Matching . . . 125
6.1 Why Are Static and Centralized Algorithms not Always Sufficient? . . . 125
6.2 Real-Time Optimization . . . 127
6.2.1 Periodic Review Method . . . 129
6.2.2 Continuous Review Method . . . 134

6.3 Distributed Control 142
6.3.1 Multi-agent Systems 142
6.3.2 Interaction Protocols 146
6.4 The "AI" Challenges (Artificial Intelligence; Analytics and Informatics) 153
6.4.1 Artificial Intelligence 153
6.4.2 Analytics and Informatics 157
6.5 Summary 159
References 163

7 Extended Examples of Best Matching 167
7.1 Understanding Through Analogy 167
7.2 E_1: Collaborative Supply Networks (M: $1/RC, RS/-, OS/DI$) 170
7.2.1 Mathematical Formulation 172
7.2.2 D+: Task Administration Protocol 174
7.2.3 Measured Impact 179
7.2.4 Discussion 180
7.3 E_2: Collaborative Assembly Lines (M: $1/RC, PR, RS/---, GP/DI$) 182
7.3.1 Mathematical Formulation 184
7.3.2 D+: Collaborative Multi-agent System 188
7.3.3 Measured Impact 192
7.3.4 Discussion 192
7.4 E_3: Clustering with Interdependent Preferences (M: $1/RC, IP/+, OS/ES$) 194
7.4.1 Optimal Clustering: Genetic Algorithm (GA) 196
7.4.2 D+: Association/Dissociation 199
7.4.3 Measured Impact 201
7.4.4 Discussion 201
7.5 E_4: Collaborative Service Enterprises ($1{:}M{:}M/RC, RS/+-, WS$) 202
7.5.1 Mathematical Formulation 204
7.5.2 Optimization: Tabu Search 206
7.5.3 Measured Impact 210
7.6 Summary 211
Appendix 1: Notation 213
References 217

8 Frontiers in Best Matching 221
8.1 Emerging Technologies Dealing with Best Matching 221
8.1.1 Connected Robots 222
8.1.2 Cloud Manufacturing 222
8.2 Technical Challenges of Best Matching 223
8.2.1 Efficient Computation and Communication 223
8.2.2 Conflict and Error Detection and Prevention 224

8.2.3 Incentives for Collaboration 224
8.2.4 Data Availability and Reliability 225
8.3 Summary 226
References 228

Erratum to: Best Matching Theory & Applications E1

Index 229

Chapter 1
Introduction: Best Matching and Best Match

Abstract This chapter presents the fundamentals of best matching in distributed systems of operations and decisions, along with several examples and illustrations. The purpose is to present the basic definitions of best matching in the context of distributed manufacturing, supply, and service systems, highlight its impact on the competitive performance of such systems, and outline the structure and objectives of this book.

1.1 What Is Best Matching?

Matching is an intrinsic process in any natural or artificial system that involves *interaction*. In games—we match cards, geometrical toy shapes, and capitals to states; in cooking—ingredients to recipes; and in everyday life—solutions to problems, remedies to pains, and tools to tasks. *Best matching*, finding the best match and doing so in the best way is often associated with survival. We all exercise it in our everyday lives, and aspire to be good at it—finding the pairing sock in the morning, the fastest or safest route to school, the most reliable business partner, the most appealing place to work, the right person to marry, and the missing piece of a puzzle. Perhaps we do not find the best match through the best, most efficient, or fastest process, but we still look for the preferred, best match, sometimes compromising with "a bird in the hand". Best matching is also found in other species—finding the shortest path between food and nest (e.g., ants), the proper role for each member of the colony (e.g., forager/nurse wasps), the most strategic place to build a nest (e.g., spiders), the most dependable member of the flock to follow (e.g., birds), and the fittest mate. These are a few out of plentiful instances of best matching in nature. Artificial systems also deal with similar, industrial and business matching processes—matching bolts and nuts (selective assembly), suppliers, distributors, retailers, and customers (enterprise network selective design), jobs and machines/computers (scheduling), nurses and shifts (service assignment), vehicles and routes (transportation planning), sensors,

M. Moghaddam and S.Y. Nof, *Best Matching Theory & Applications*, Automation, Collaboration, & E-Services 3, DOI 10.1007/978-3-319-46070-3_1

locations, and positions (sensor network design), interns and factories/hospitals (recruitment), robots and teams (team formation).

Best matching is essentially based on the interaction of individuals with each other and with their environment. Those interactions, in turn, define their desirability or *preferences* for each other along with certain *conditions*. The types of matching preferences may considerably vary, depending on the context, e.g., dimensional tolerance (bolts and nuts); lead-time/cost/quality (suppliers and customers); tardiness/makespan (jobs and machines); delivery time/cost (vehicles and routes); communication cost/energy consumption (sensors and locations); employee/employer satisfaction (interns and factories); conflict rate/resilience (robots and teams). Likewise, the conditions that influence a best matching process are context-specific, e.g., limitation on the number of interns that a factory or a hospital can admit; precedence relations among a set of jobs allocated to a single computer; lateral collaboration among suppliers through demand and capacity sharing; interpersonal biases, emotions, and relational messages among members of a social network.

Oxford dictionaries define matching as [the process of] "corresponding or causing to correspond in some essential respect", "making or being harmonious", or "being equal to (something) in quality or strength". Given that, best matching is defined as

Best Matching: The process of finding the best match between the elements of two or more sets, considering certain conditions and criteria.

The best matching process may be *static* or *dynamic*. For instance, location of a set of warehouses is regarded as a relatively static best matching process (warehouses and locations), whereas a multi-robot system may involve highly dynamic best matching processes (robots and teams) with real-time reconfiguration of the

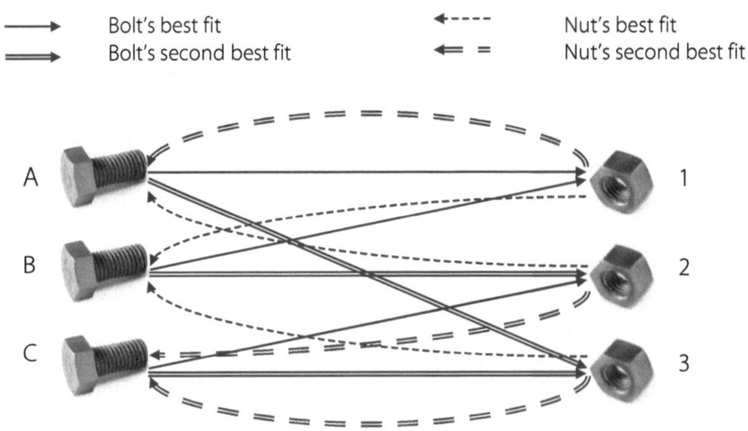

Fig. 1.1 Matching bolts and nuts. Bolts/nuts deviate within their predefined dimensional tolerances, their matching to nuts/bolts with the same "type" of deviation may result in a higher quality product. Considering *overall quality* as the matching criterion, the best combination of bolts and nuts is {A-2; B-1; C-3} (Velásquez and Nof 2008)

teams. Figure 1.1 shows a classic example of best matching processes in selective assembly.

Mismatch between individual entities of distributed systems is a challenging problem. Although a system with "mismatched individuals" may not necessarily collapse, it is certainly outperformed by an equivalent system with "best-matched individuals", in terms of a variety of critical criteria such as time, cost, quality, flexibility, and stability. The key challenge, therefore, is to identify those matching processes in a given distributed system, specify, structure, and formulate them in a systematic manner, and develop, test, and validate algorithms and protocols for solving them in an efficient manner. Best matching is an important problem that roots in almost every natural and artificial system. Best matching, however, is not a new problem. It was formally introduced by Votaw and Orden (1952) as the *assignment problem*, and has been extensively studied for more than six decades. Several models and algorithms were developed to find an optimal match, led by pioneering works such as *Hungarian algorithm* by Kuhn (1955) and *deferred acceptance algorithm* by Gale and Shapley (1962). In 2012, Lloyd Shapley and Alvin Roth were awarded the Nobel Memorial Prize in Economic Science for their work on market design and matching theory. Thousands of articles, chapters, and books have been published in this well-established area. The goal of this book is not to rehash all these materials. We are not reinventing the wheel. The goal of this book is to elaborate on the fundamentals of best matching in distributed and collaborative systems by providing solid frameworks for

- Methodical analysis of various best matching processes from different dimensions, and identification of new classes of the problem.
- Comprehensive taxonomy, analogical and structural comparison between different best matching problems and processes.
- Systematic identification of systems' hierarchy and nature of interactions as well as distribution of decision-making and control functions among distributed individuals.
- Practical formulation of solutions based on a resourceful library of tools—best matching algorithms and protocols.

The book covers diverse examples of best matching in different production, manufacturing, and service systems, and provides guidelines for applications and for future research in both basic and applied directions. Compared with related references, this book is focused more on the systems engineering and applied operations research categories, and it is designed for both academics and practitioners.

1.2 Definitions and Scope

Any system that involves interactions among its distributed individuals may undergo certain best matching processes. From a systematic perspective, matching involves three general elements (Fig. 1.2): The matching goal, representing the motivation and need for matching; the process logic of the desired matching, specifying how to execute the matching and derive the resulting match; and the resulting match itself. Although some problems may be solved satisfactorily with any matching process and any matching result, the term "best matching" here implies attention the finding better matches, possibly through better processes. Consider as an example—besides the examples discussed in the previous section—a manufacturing system with multiple layers and various cyber-physical components or *collaborative units* (Fig. 1.3):

- *The AS/RS* (*Automated Storage/Retrieval System*) must identify the best storage locations for different parts/products. AS/RS is typically used in applications with high volumes/rates of arrival and retrieval, strict capacity limits, and/or critical accuracy requirements. In such systems, best matching can play a significant role in enhancing efficiency and productivity.
- *The process robots* must be assigned the most suitable sets of tasks. Automated Guided Vehicles (AGVs), as an example of process robots, are responsible for carrying materials around the shop floor or warehouse. When multiple AGVs are working in the same facility, it is necessary to define *who does what*, *how*, and *when*, in order to minimize processing times and prevent conflicts (e.g., collision). This can be done through dynamic best matching between materials and AGVs.
- *The facility sensors* must find the best peer sensors based on their network configuration/communication protocols to minimize energy consumption. Facility sensor networks are formed in different manners (e.g., single-multi-hop with/without clustering) and follow certain communication protocols (e.g., point-to-point; flooding; gossiping), in an attempt to minimize energy consumption through sending/receiving messages. One of the foremost processes in optimal configuration of sensor network as well as development of efficient communication protocols is best matching (between, e.g., sensors; regions; clusters; cluster heads; base stations).

Fig. 1.2 Three general elements of matching: the goal, the process logic, and the result

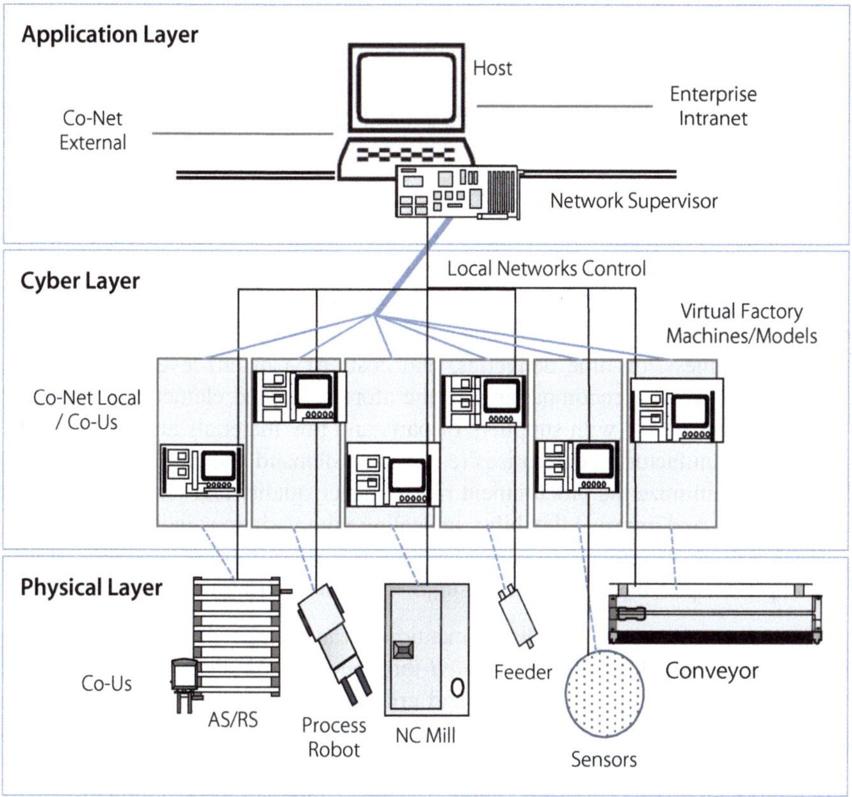

Fig. 1.3 A cyber-physical manufacturing environment

- *The feeder lines* must rearrange/combine different component parts based on their similarity/affinity/dimensional tolerance. The bolt-nut best matching process (Fig. 1.1) is a good example to show the significant impact of best matching on selective assembly. Matching the component parts that arrive in the primary assembly line from different feeder lines (based on, e.g., tolerance) can remarkably improve the quality of final products (even in presence of manufacturing deficiencies and inaccuracies), and reduce the need for rework.
- *The NC (Numerical Control) milling machines* must define the best tools with respect to the assigned jobs. Milling machines use various rotary cutters to remove materials from the surface of a work-piece. The rotary cutters differ in shape, size, material type, flutes/teeth, helix angle, coating, shank, etc., each suitable for different types of work-pieces. The wide range of choices and features of rotary cutters provide challenges for the NC program to match the best tool to each work-piece (based on the features of both). The problem becomes even more challenging when multiple NC milling machines collaborate by sharing tools and the number/diversity of the tools is limited.

- *The virtual factory manager* must identify the best virtual machine(s)/model(s) with respect to their workload/capabilities as well as the command type. The main purpose of virtual factories is to enable innovative and cost-effective production through adaptive design, virtual modeling and simulation, automated monitoring of products, processes, and factories, and knowledge integration. In this context, dynamic and optimal matching of modeling, analysis, and decision-making tasks to distributed resources (e.g., computer agents; programs; machines) with various workloads and limited capacities can enhance the productivity and flexibility of the cyber, and, in turn, physical layers of the manufacturing system.
- *The manufacturing site* must categorize its suppliers based on their quality, trustworthiness, on-time deliveries, and cost. At a macro level, the manufacturing enterprise—encompassing all the aforementioned elements—must interact and collaborate with suppliers of parts and raw materials and perhaps other similar manufacturing enterprises (e.g., lateral demand and capacity sharing). In order to minimize the procurement risk, enhance quality and service level to the customers, and improve flexibility in dealing with variations in demand and spot market price, the enterprise must identify the best portfolio of suppliers for each component part based on certain supplier selection criterion/criteria.

The common property of all the mentioned (and similar) systems is that they involve multiple individuals, *distributed* throughout the system, looking for their best match under certain conditions and criteria. Depending on the nature of the system and self- versus team-orientation, the individuals may display one of the following two fundamental behaviors during the best matching process: *collaboration* or *competition*. The following subsections elaborate more on the distributed nature of the emerging production, manufacturing, and service systems and their respective best matching processes, as well as the implied behaviors of individuals in different systems and cases.

1.2.1 Distributed Systems

The term *distributed system* is typically known as a concept in Computer Science and Engineering, which refers to a software system with components located in networked computers/computational nodes that communicate and coordinate their tasks through certain protocols. Our definition of distributed systems, however, is not limited to distributed computing and operating systems. It encompasses all large-scale systems, or *systems of systems*, with internetworked cyber/physical elements that interact and cooperate to realize the goals at system-wide and local levels (Colombo et al. 2010). The growing dominance of the Internet in almost every system, the advent of e-Work, and evolution of Information and Communication Technologies (ICT) together with the increasing complexity and dynamicity of emerging e-Systems have transformed classic monolithic systems to distributed networks of agents (Nof et al. 2015, Chap. 4). The *co-evolution* of systems and their supportive technologies (see, e.g., Fig. 1.4) has brought up emerging design criteria such as agility, scalability, resilience, and adaptability (Putnik et al. 2013), all rooted in and enabled by the *intelligence*, *autonomy*, and *collaborability* of individuals (Moghaddam and Nof 2015a).

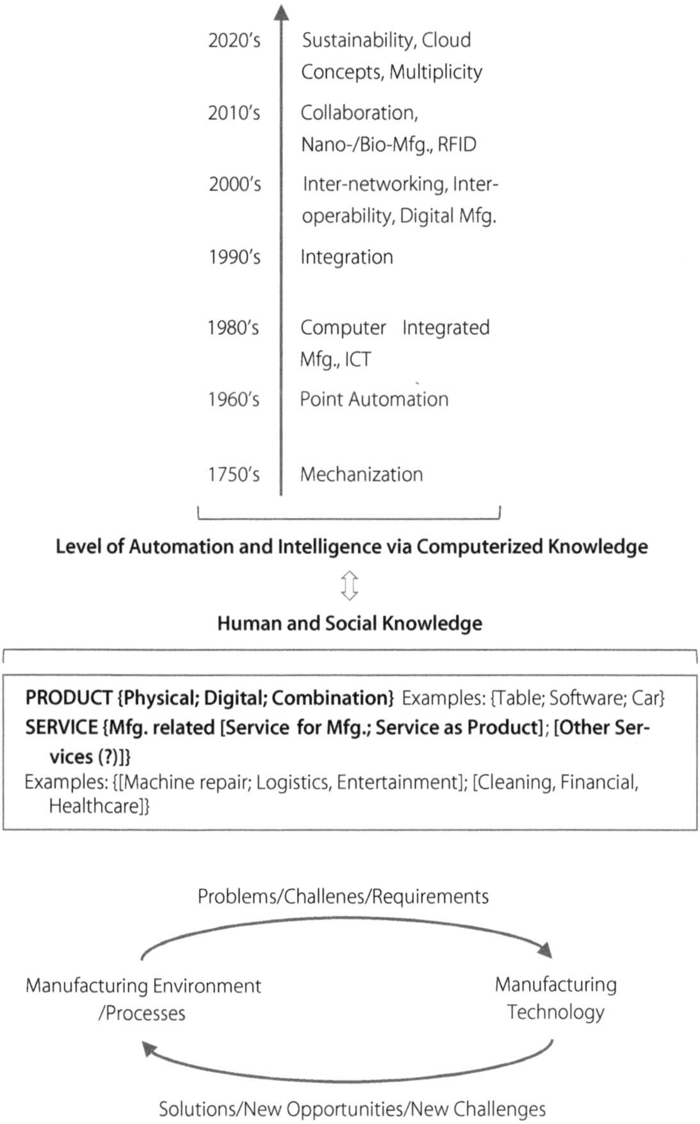

Fig. 1.4 Manufacturing, intelligence, knowledge, and services—evolution of concepts, challenges, and technologies (adapted from Nof 2013)

With the above definitions, it can be argued that any type of best matching processes is essentially attributed to a distributed system of elements categorized into at least two distinct sets. Although the details may be widely different, a considerable analogy exists between the general structures of the systems that fit into our definition of distributed systems. The range of applications, therefore, can be as wide as the variety of such systems—from a pack of bolts and nuts with no intelligence to a team of super-intelligent robots, from a small network of micro-/nano-sensors to large-scale global enterprise networks. Agent-based technologies have enabled the distribution of *intelligence*, *autonomy*, and *decision-making authority*—and thus, *control*—among a collection of homogenous/heterogeneous and networked agents (Leitão 2009; Barbati et al. 2012). Application areas include (but are not limited to) intelligent manufacturing, agriculture, security, defense, transportation, supply network design, air traffic control, and healthcare (Jennings 1998; Nof et al. 2015, Chap. 4). It is shown throughout the book that best matching is a key enabler of the emerging design criteria for distributed systems, is enhanced by intelligent and autonomous behaviors/capabilities of individuals, and has mutual influence on the collaboration mechanisms in distributed systems. These topics are discussed in more detail in the next chapters, especially Chaps. 4 and 6.

1.2.2 Collaboration Versus Competition

Collaboration and competition influence matching processes and the resulting match in several ways. Collaboration refers to interactive processes where "*all involved parties share information, resources and responsibilities to jointly plan, implement, and assess the set of activities required to achieve a common goal, thus jointly creating value added*" (Nof 2007). Collaboration is now emerging as a new branch of science, a subcategory of *interaction science*, and its scope is as broad as any natural or artificial distributed system that involves, to some extent, the aforementioned process of sharing. Collaboration enables powerful augmentation of any system to work, sustain, and evolve. It is a universal principle. Collaboration is a necessary mechanism for the survival and evolution of any natural and artificial organization, from genomes, multi-cellular organisms, social insects, and human societies to networks of sensors, computers, robots, and global enterprises. Collaborative processes may vary, depending on the level, intensity, and type (mandatory/optional/concurrent) of collaboration (Nof et al. 2015, Chap. 2). The primary challenge, therefore, is how to understand, engineer, improve, systematize, and automate collaboration processes and generate knowledge about them, in distributed systems of humans, machines, robots, software agents, etc.

According to its definition above, collaboration has three different implications in the context of best matching:

(1) Sharing information;
(2) Sharing resources;
(3) Sharing responsibilities.

The first case is general—individuals have to share their local information (e.g., preferences; capacity limits) with others during the best matching process. The second case refers to sharing physical/virtual resources among the individuals of the same set or different sets during the best matching process; e.g., customer allocation —demand and capacity sharing among suppliers; assembly line balancing—tool sharing among workstations. In the first two cases, sharing information and resources, individuals do not necessarily seek to satisfy common objectives—they can even be *competitors* and still share information and resources (e.g., Subaru of Indiana Automotive, Inc., building Toyota Camry cars alongside Subaru cars). What makes the difference is the third case, *sharing responsibilities*.

Collaboration is known as a critical mechanism of survival and evolution in nature—although the notion of natural selection implies competition, Nowak's rules for the evolution of collaboration in nature (Nowak 2006) indicate that selfish replicators forgo some of their reproductive potential to help one another (Fig. 1.5). In artificial systems, however, competition is generally an inevitable behavior—companies compete for customers/markets; students compete for internship positions; negotiation agents compete for products/services; hospitals compete for good doctors; orders compete for resources. It is all about *incentives* for collaboration. Some researchers and practitioners believe that collaboration, by itself, may not guarantee desired success—why should we share and not compete? What are the role and optimal level of collaboration in ensuring effective competitiveness and winning? While we describe and explain in this book techniques and tools to analyze and address these questions, the answers to them are beyond the scope of this book. (See Nof et al. 2015, Chap. 11, for more details.) Relevant here is the trade-off between self-orientation versus group-orientation, which categorizes best matching processes into the following classes:

- *Collaborative*. All individuals pursue a set of global objectives that are in line with their local objectives.
- *Competitive*. Individuals pursue local objectives that may be conflicting with the given global objectives.

1.3 Best Matching in Practice

Applications of best matching, as discussed earlier, are widely diverse, involving different production, manufacturing, and service industries. Various scheduling (e.g., machine; computational grids; workforce planning; batching; dispatching; load balancing), transportation and routing (e.g., air traffic control; train decoupling; aeromedical routing), telecommunication (e.g., grid computing; sensor networks;

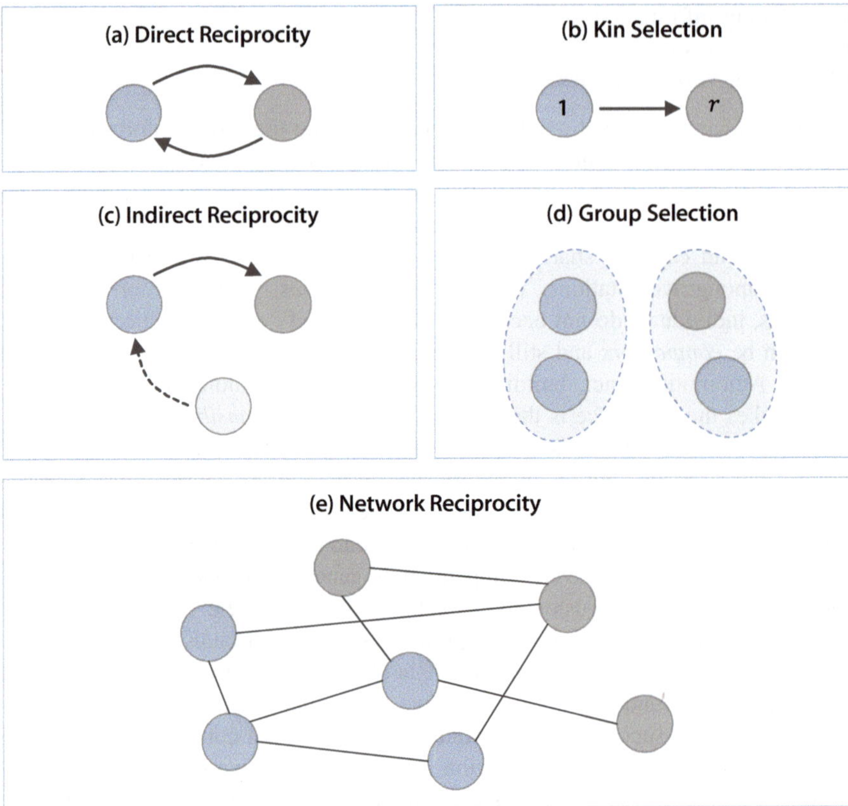

Fig. 1.5 Evolution of collaboration in nature. **a** Willingness to collaborate depends on repeated encounters between the same two individuals. **b** The donor and the recipient of a collaborative act are genetic relatives. **c** Reputation-based—a helpful individual is more likely to receive help. **d** Competition is not only between individuals but also between groups (depends on size and number of groups). **e** Clusters of collaborators are more likely to outcompete against defectors (extended from Nowak 2006)

swarm robotics), production planning (e.g., batch loading, group technology, order selection, lot scheduling), supply network design and logistics (e.g., supplier selection; warehousing; demand partitioning; sourcing; market clearing) processes can be regarded as instances of best matching in practice. Measures for evaluating best matching and best match have been applied in research and practice; for example, *economic value of match*, *economic cost of match*, *random fit index*, and *best fit index*, for individuals and for the entire system (Kang 1994; Chiam and Nof 2005; Velásquez and Nof 2007, 2008).

A basic, typical (not necessarily simple) best matching is the matching of two sets of elements. Hence, it is called two-sided matching process. Examples are (Cattrysse and Van Wassenhove 1992) fixed-charge location problems, grouping and loading in flexible manufacturing systems, scheduling projects, storage allocation, designing a communication network, assigning jobs to computers, scheduling variable length TV commercials, and assigning ships to overhaul facilities (see Table 1.1 for more examples). Classic examples of matching

Table 1.1 Real-life examples of two-sided best matching (expanded from Öncan 2007)

Problem	Set 1	Set 2	Reference
Selective assembly	Tolerances	Parts	Buckingham (1920)
Automotive assembly	Cylinders	Pistons	Mansoor (1961)
Workforce scheduling	Manpower	Sections	Littschwager and Tcheng (1967)
Location-allocation	Suppliers	Consumers	Ross and Solland (1977)
Maximal covering	Sites	Customers	Klastorin (1979)
Resource scheduling	Institutions	Activities	Zimokha and Rubinstein (1988)
Capacity planning	Periods	Batches	Mazzola et al. (1989)
Group formation	Machines	Parts	Shtub (1989)
Database partitioning	Processors	Partitions	Boffey (1989)
Project management	Labors	Jobs	Drexl (1991)
Storage assignment	Locations	Items	Lee (1992)
Components integration	Components	Sets/teams	Nof (1994a, b)
Disposal assignment	Disposers	Removers	Campbell and Langevin (1995)
Machine assignment	Cells	Machines	Cheng et al. (1996)
Parts assembly	Suppliers	Assemblies	Nof and Kang (1996)
Dairy farm allocation	Factories	Suppliers	Foulds and Wilson (1999)
Dynamic order selection	Periods	Orders	Lee and Kim (1998)
Stock management	Demands	Stocks	Privault and Herault (1998)
Aeromedical routing	Flights	Patients	Ruland (1999)
Vehicle routing	Vehicles	Cities	Baker and Sheasby (1999)
Land use allocation	Activities	Lands	Cromley and Hanink (1999)
Telescope scheduling	Intervals	Activities	Nowakovski et al. (1999)
Production planning	Agents	Tasks	LeBlanc et al. (1999)
Third party routing	Depots	Customers	Jalisi and Cheddad (2000)
Market clearing	Asks	Bids	Kalagnanam (2001)
Capacitated clustering	Seeds	Customers	Shieh and May (2001)
Worker allocation	Depts.	Workers	Campbell and Diaby (2002)
Single egress selection	Edge links	Prefixes	Bressoud et al. (2003)
Power management	Voltage	Tasks	Yu and Prasanna (2003)
Single sourcing	Warehouses	Customers	Freling et al. (2003)
Wireless network design	Base-station	Terminals	Barbas and Marin (2004)
Demand partitioning	Facilities	Products	Benjafaar et al. (2004)
Load balancing	Machines	Tasks	Harvey et al. (2006)
Supplier selection	Suppliers	Components	Velasquez and Nof (2008)
Service allocation	Branches	Markets	Moghaddam et al. (2015)
Nano-sensor fabrication	DNA	Silicon plate	Mori et al. (2015)
Micro-manufacturing	Components	Devices	Agarwal and Livermore (2016)
Campus placements	Skills	Institutions	Upadhayay and Vrat (2016)

problems with more sets involved include time-based allocation of bus drivers to routes (Carraresi and Gallo 1984), and medical residents to rotations (Franz and Miller 1993), followed by more advanced examples such as design of cyber-physical distributed systems or large-scale networks of multi-national corporations (Moghaddam and Nof 2015b). Following Chaps. 1–8 of the book provide a more elaborate and comprehensive review of best matching processes in various industrial and service sectors.

1.4 Summary

The focus of this book is on both the theory and applications of best matching. Thousands of best matching problems have been identified and solved in the past. Several articles and books have analyzed the complexity of different problem instances and provided efficient algorithms and protocols for solving them. The goal of this book is to present a theoretical foundation for best matching along with necessary tools and techniques for *identifying*, *classifying*, and *solving* different instances of best matching problems in practice—a "best matching manual" with a balanced combination of theory and practice. It provides essential algorithms and protocols for the design of best matching processes, yet opens and leaves some theoretical discussions (e.g., on complexity theory) to relevant and already published works (e.g., Burkard et al. 2009). Although the examples discussed throughout this book are mainly related to manufacturing and production environments, the reader can generalize, through the examples discussed in this chapter, the role and impact of best matching in a wide range of situations. The models and tools presented in this book are applicable in other settings as well.

This book is divided into three parts. The first part presents the theoretical foundation for identification and classification of various best matching instances. Chapter 2 presents the recently developed *PRISM taxonomy of best matching*, which characterizes and formalizes best matching processes based on 3 + 1 dimensions: Sets (D1); Conditions (D2); Criteria (D3); Time or Progression (D+). This conceptual framework uncovers remarkable opportunities and challenges for future research in this area: While best matching is well established from the perspectives of D1 and D3, it has not received similar attention with respect to D2 and D+. Following the PRISM taxonomy, Chap. 3 presents basic mathematical formulations, in the form of Mixed-Integer Programs (MIPs), for different instances of best matching problems, along with several examples and illustrations.

The second part of the book presents fundamental mechanisms for handling different instances of best matching in classic monolithic and distributed systems. Chapter 4 justifies the need for algorithms and protocols based on the structures of decision-making hierarchies, from purely centralized to totally distributed and interactive behaviors of individuals (i.e., collaborative; competitive; hybrid). Chapter 5 presents major exact, heuristic, and metaheuristic best matching algorithms. Chapter 6 covers the basics of best matching protocols by brief introduction of agent-based systems and various control architectures (i.e., centralized; hierarchical; heterarchical; distributed); followed by major interaction protocols (e.g., negotiation protocols; contract net protocols; task administration protocols), and D+ related principles of CCT (Collaborative Control Theory), the principles of *association-dissociation* and *emergent lines of collaboration.* Several examples are presented for better illustration and understanding of the presented concepts.

The third part of the book discusses various applications of best matching processes in production and manufacturing (e.g., scheduling; material handling; shop floor control; warehouse operations; supply network design and coordination), and service (e.g., healthcare; transportation; power systems; agriculture; defense) industries (Chap. 7). It concludes the book's discussions by presenting major trends and guidelines for basic and applied research in the area of best matching in the future (Chap. 8).

Problems

1.1. Identify the three general elements of matching in the following examples:

- (a) Playing Sudoku.
- (b) Playing a soccer match.
- (c) Assigning security personnel in an airport.
- (d) Solving a jigsaw puzzle.
- (e) Recruiting a machinist to a production line.
- (f) Selecting a supplier, e.g., of food, education, healthcare, transportation.

1.2. In Problem 1.1, which cases may result in multiple matches, of which only one can be considered "best"?

1.3. In Problem 1.1, does the matching process influence the resulting match? Explain how.

1.4. Can a bad matching process yield a best match? Illustrate for the cases in Problem 1.1.

1.5. Does a best (optimal) matching process guarantee a best match? Illustrate for the cases in Problem 1.1.

1.6. Answer Problems 1.1–1.5 for some of the matching cases in Fig. 1.3.

1.7. Select your own example of a matching case, and answer Problems 1.1–1.5 for it.

1.8. Explain why (or under what conditions) each of the following cases implies no requirement for matching, requirement for some matching, or requirement for best matching:

- (a) Basketball league game.
- (b) New product launch.
- (c) Equipment repair.
- (d) Furniture assembly.
- (e) Homework assignment.
- (f) Financial investment decision.
- (g) Meal preparation.

(h) Information search.
(i) Painting a picture.
(j) Contacting a customer.
(k) Buying a gift.
(l) Setting an alarm clock.

1.9. For cases in Problem 1.8 that require matching (some or best), explain whether it requires also a best resulting match.

1.10. For each case in Problem 1.8 that requires matching (some or best), explain whether it requires a single matching process execution, or multiple and repeated matching executions.

1.11. Describe three examples where the main objective (what really matters) is:

(a) The matching process itself.
(b) The best resulting match.
(c) Both the matching process and the best resulting match.

1.12. Identify in the examples of the above problems:

(a) Distributed versus centralized matching.
(b) Competitive versus collaborative matching.
(c) Information sharing processes for the purpose of best matching.
(d) Resource sharing processes for the purpose of best matching.
(e) Responsibilities sharing processes for the purpose of best matching.

1.13. For each of the five cases of evolution of collaboration in nature (Fig. 1.5), illustrate one problem of best matching, and answer Problem 1.12 specifically for it.

1.14. Compare and discuss the similarities and differences between best matching problems, assignment problems, scheduling problems, and resource allocation problems.

1.15. Review the examples of two-sided matching in Table 1.1, and describe:

(a) An example of one-sided matching.
(b) An example of three-sided matching.

1.16. Using examples from Table 1.1, discuss several differences between collaborative and competitive best matching.

References

Agarwal, G., & Livermore, C. (2016). Shape-selective assembly of anisotropic, deformable microcomponents using bottom-up micro-manufacturing. *Micromachines, 7*, 68–85.

Baker, M. B., & Sheasby, J. (1999). Extensions to the generalized assignment heuristic for vehicle routing. *European Journal of Operational Research, 119*, 147–157.

Barbas, J., & Marin, A. (2004). Maximal covering code multiplexing access telecommunication networks. *European Journal of Operational Research, 159*, 219–238.

Barbati, M., Bruno, G., & Genovese, A. (2012). Applications of agent-based models for optimization problems: A literature review. *Expert Systems with Applications, 39*, 6020–6028.

Benjaafer, S., El-Hafsi, M., & de Vericourt, F. (2004). Demand allocation in multiple-product, multiple-facility, make-to-stock systems. *Management Science, 50*, 1431–1448.

Boffey, T. B. (1989). Location problems arising in computer networks. *Journal of the Operational Research Society, 40*, 347–354.

Bressoud, T. C., Rastogi, R., & Smith, M. A. (2003). Optimal configuration for BGP route selection. In *Proceedings of IEEE INFOCOM' 2003, 2nd Annual Joint Conference on the IEEE Computer and Communications Societies, San Francisco* (Vol. 2, pp. 916–926).

Buckingham, E. (1920). Manufacturing for selective assembly. *Machinery, 27*, 142–143.

Burkard, R. E., Dell'Amico, M., & Martello, S. (2009). *Assignment Problems*. Siam, ISBN 978-1-611972-22-1.

Campbell, G. M., & Diaby, M. (2002). Development and evaluation of an assignment heuristic for allocating cross-trained workers. *European Journal of Operational Research, 138*, 9–20.

Campbell, J. F., & Langevin, A. (1995). The snow disposal assignment problem. *Journal of the Operational Research Society, 46*, 919–929.

Carraresi, P., & Gallo, G. (1984). A multi-level bottleneck assignment approach to the bus drivers-rostering problem. *European Journal of Operational Research, 16*, 163–173.

Cattrysse, D. G., & Van Wassenhove, L. N. (1992). A survey of algorithms for the generalized assignment problem. *European Journal of Operational Research, 60*, 260–272.

Cheng, C. H., Goh, C.-H., & Lee, A. (1996). Solving the generalized machine assignment problem in group technology. *Journal of the Operational Research Society, 47*, 794–802.

Chiam, T. C., & Nof, S. Y. (2005). Modeling and application of the best matching protocol. *PRISM Research Memorandum No. 2005-P1*, Purdue University, West Lafayette, IN.

Colombo, A. W., Karnouskos, S., & Mendes, J. M. (2010). Factory of the future: A service-oriented system of modular, dynamic reconfigurable and collaborative systems artificial intelligence techniques for networked manufacturing enterprises management. *Springer Series in Advanced Manufacturing*, 459–481.

Cromley, R. G., & Hanink, D. M. (1999). Coupling land use allocation models with raster GIS. *Journal of Geographical Systems, 1*, 137–153.

Drexl, A. (1991). Scheduling of project networks by job assignment. *Management Science, 37*, 1590–1602.

Foulds, L. R., & Wilson, J. M. (1999). On an assignment problem with side constraints. *Computers & Industrial Engineering, 37*, 847–858.

Franz, L. S., & Miller, J. L. (1993). Scheduling medical residents to rotations, solving the large-scale multi-period staff assignment problem. *Operations Research, 41*, 269–279.

Freling, R., Romeijn, H. E., Romero-Morales, D., & Wagelmans, A. P. M. (2003). A branch-and-price algorithm for the multi-period single-source problem. *Operations Research, 51*, 922–939.

Gale, D., & Shapley, L. S. (1962). College admission and stability of marriage. *The American Mathematical Monthly, 69*, 9–15.

Harvey, N. J. A., Ladner, R. E., Lovasz, L., & Tamir, T. (2006). Semi-matchings for bipartite graphs and load balancing. *Journal of Algorithms, 59*, 53–78.

Jalisi, Q. W. Z., & Cheddad, H. (2000). Third party transportation, a case study. *International Journal of Industrial Engineering, Theory Applications Practice, 7*, 348–351.

Jennings, N. R., Sycara, K., & Wooldridge, M. (1998). A roadmap of agent research and development. *Autonomous Agents and Multi-Agent Systems, 1*, 7–38.

Kalagnanam, J. R., Davenport, A. J., & Lee, H. S. (2001). Computational aspects of clearing continuous call double auctions with assignment constraints and indivisible demand. *Electronic Commerce Research, 1*, 221–238.

Kang, H. (1994). Development of information exchange protocols for distributed inspection integration. *M.S. Thesis*. School of Industrial Engineering, Purdue University.

Klastorin, T. D. (1979). On the maximal covering location problem and the generalized assignment problem. *Management Science, 25*, 107–112.

Kuhn, H. W. (1955). The Hungarian method for the assignment problem. *Naval Research Logistics Quarterly, 2*, 83–97.

LeBlanc, L. J., Shtub, A., & Anandalingam, G. (1999). Formulating and solving production planning problems. *European Journal of Operational Research, 112*, 54–80.

Lee, M.-K. (1992). A storage assignment policy in a man-on-board automated storage/retrieval system. *International Journal of Production Research, 30*, 2281–2292.

Lee, D.-H., & Kim, Y.-D. (1998). A multi-period order selection problem in flexible manufacturing systems. *Journal of the Operational Research Society, 49*, 278–287.

Leitão, P. (2009). Agent-based distributed manufacturing control: A state-of-the-art survey. *Engineering Applications of Artificial Intelligence, 22*, 979–991.

Littschwager, J. M., & Tcheng, T. H. (1967). Solution of a large-scale forest scheduling problem by linear programming decomposition. *Journal of Forestry, 65*, 644–646.

Mansoor, E. M. (1961). Selective assembly—Its analysis and applications. *International Journal of Production Research, 1*, 13–24.

Mazzola, J., Neebe, A., & Dunn, C. (1989). Production planning of a flexible manufacturing system in a material requirements planning environment. *International Journal of Flexible Manufacturing Systems, 1*, 115–142.

Moghaddam, M., & Nof, S. Y. (2015a). The collaborative factory of the future. *International Journal of Computer Integrated Manufacturing*. doi:10.1080/0951192X.2015.1066034

Moghaddam, M., & Nof, S. Y. (2015b). Collaborative location-allocation decisions in networked v-Organizations. *Research memo, PRISM Center, Purdue University*.

Moghaddam, M., Nof, S. Y., & Menipaz, E. (2015). Design and administration of collaborative networked headquarters. *International Journal of Production Research*. doi:10.1080/00207543.2015.1125544

Mori, Y., Ma, Z., Park, S., Hirai, Y., Tsuchiya, T., & Tabata, O. (2015). *Proceedings of Transducers—2015, 18th International Conference on Solid-State Sensors, Actuators and Microsystems* (pp. 1389–1392).

Nof, S. Y. (Ed.). (1994a). *Information and collaboration models of integration*. Dordrecht: Kluwer Academic Publishers.

Nof, S. Y. (1994b). Recent developments in simulation of integrated engineering environments. In *Proceedings of SCS Symposium on AI & Computer Simulation*, Mexico City.

Nof, S. Y. (2007). Collaborative control theory for e-Work, e-Production, and e-Service. *Annual Reviews in Control, 31*, 281–292.

Nof, S. Y. (2013). Research advances in manufacturing with service-oriented e-work and production (Plenary). *IFAC Intelligent Manufacturing Systems Workshop, 11*, 251–256.

Nof, S. Y., Ceroni, J., Jeong, W., & Moghaddam, M. (2015). Revolutionizing collaboration through e-Work, e-Business, and e-Service. *Springer series in ACES, Automation, Collaboration & E-Service*. doi:10.1007/978-3-662-45777-1

Nof, S. Y., & Kang, H. (1996). Inspection-based integration protocols for best matching of components from distributed suppliers. *IE Research Memo 96-K*, Purdue University, West Lafayette, IN.

Nowak, M. A. (2006). Five rules for the evolution of cooperation. *Science, 314*, 1560–1563.

Nowakovski, J., Schwarzler, W., & Triesch, E. (1999). Using the generalized assignment problem in scheduling the ROSAT space telescope. *European Journal of Operational Research, 112*, 531–541.

Öncan, T. (2007). A survey of the generalized assignment problem and its applications. *INFOR, 45*, 123–141.

Privault, C., & Herault, L. (1998). Solving a real-world assignment problem with a metaheuristic. *Journal of Heuristic, 4*, 383–398.

Putnik, G., Sluga, A., ElMaraghy, H., Teti, R., Koren, Y., Tolio, T., et al. (2013). Scalability in manufacturing systems design and operation: State-of-the-art and future developments roadmap. *CIRP Annals—Manufacturing Technology, 62*, 751–774.

Ross, G. T., & Soland, R. M. (1977). Modeling facility location problems as generalized assignment problems. *Management Science, 24*, 345–357.

Ruland, K. S. (1999). A model for aeromedical routing and scheduling. *International Transactions in Operational Research, 87*, 57–73.

Shieh, H.-M., & May, M.-D. (2001). Solving the capacitated clustering problem with genetic algorithms. *Journal of the Chinese Institute of Industrial Engineers, 18*, 1–12.

Shtub, A. (1989). Modelling group technology cell formation as a generalized assignment problem. *International Journal of Production Research, 27*, 775–782.

Upadhayay, L., & Vrat, P. (2016). An ANP based selective assembly approach incorporating Taguchi's quality loss function to improve quality of placements in technical institutions. *TQM Journal, 28*, 112–131.

Velásquez, J. D., & Nof, S. Y. (2007). Best-matching protocol for inspected supplies over collaborative e-Work networks. In *Proceedings of ICPR-19*, Valparaiso, Chile.

Velásquez, J. D., & Nof, S. Y. (2008). A best-matching protocol for collaborative e-Work and e-Manufacturing. *International Journal of Computer Integrated Manufacturing, 21*, 943–956.

Votaw, D. F., & Orden, A. (1952). The personnel assignment problem. In Symposium on Linear Inequalities and Programming, *SCOOP 10, US Air Force* (pp. 155–163).

Yu, Y., & Prasanna, V. K. (2003). Resource allocation for independent real-time tasks in heterogeneous systems for energy minimization. *Journal of Information Science and Engineering, 19*, 433–449.

Zimokha, V. A., & Rubinstein, M. I. (1988). R & D planning and the generalized assignment problem. *Automation and Remote Control, 49*, 484–492.

Chapter 2
The PRISM Taxonomy of Best Matching

Abstract This chapter presents the notion of the PRISM taxonomy of best matching along with its 3 + 1 dimensions, as a taxonomic framework for systematic identification and specification of best matching processes in various and unrelated domains. Several examples from manufacturing, supply, and service systems are presented to illustrate and exemplify the features and applications of this framework. The purpose is to enable understanding and problem solving capabilities through *analogical reasoning*—comparative analysis of different matching processes by means of the PRISM taxonomy.

2.1 Framework

The first chapter presented several examples of best matching in a wide variety of contexts and application domains. It was shown that while certain problems may seem at first to be completely unrelated to matching, they could actually be formalized as certain classes of best matching. This chapter introduces a new fundamental framework for identification and classification of different best matching processes. The goal is to provide a robust collection of modeling, design, and optimization tools for different classes of best matching problems, enabled by *analogical reasoning*—solving a problem based on the solution to its "analogous story". Before describing the mechanics of the taxonomic framework, it is worthwhile to briefly describe the notion of analogical reasoning and the way it justifies the need for such a framework in the context of best matching.

Analogical reasoning refers to a thinking process that relies on analogy, i.e., "a comparison between two things, typically for the purpose of explanation or clarification" (Oxford dictionaries). Consider *Duncker's radiation problem* (Duncker 1945) as an example:

> Suppose you are a doctor faced with a patient who has a malignant tumor in his stomach. It is impossible to operate on the patient, but unless the tumor is destroyed, the patient will die. There is a kind of ray that can be used to destroy the tumor. If the rays reach the tumor

M. Moghaddam and S.Y. Nof, *Best Matching Theory & Applications*, Automation, Collaboration, & E-Services 3, DOI 10.1007/978-3-319-46070-3_2

all at once at a sufficiently high intensity, the tumor will be destroyed. Unfortunately, at this intensity the healthy tissue that the rays pass through on the way to the tumor will also be destroyed. At lower intensities, the rays are harmless to healthy tissue, but they will not affect the tumor, either. What type of procedure might be used to destroy the tumor with the rays and at the same time avoid destroying the healthy tissue?

A selected group of people was asked to solve the radiation problem—few people did. The reason was the *ill* definition of the problem, which essentially required an "analogy" from a remote domain to trigger creativity (Gick and Holyoak 1980). Experiments revealed that if the definition of the problem preceded a story with an analogous *convergence solution*, the majority of people would be able to solve it. The analogous story was named *attack-dispersion*:

A small country was ruled from a strong fortress by a dictator. The fortress was situated in the middle of the country, surrounded by farms and villages. Many roads led to the fortress through the countryside. A rebel general vowed to capture the fortress. The general knew that an attack by his entire army would capture the fortress. He gathered his army at the head of one of the roads, ready to launch a full-scale direct attack. However, the general then learned that the dictator had planted mines on each of the roads. The mines were set so that small bodies of men could pass over them safely, since the dictator needed to move his troops and workers to and from the fortress. However, any large force would detonate the mines. It therefore seemed impossible to capture the fortress. However, the general devised a simple plan. He divided his armies into small groups and dispatched each group to the head of a different road. When all was ready, he gave the signal and each group marched down a different road. Each group continued down its road so that the entire army arrived together at the fortress at the same time. In this way, the general captured the fortress and overthrew the dictator.

Analogical reasoning is the foundation for creativity in humans and even in animals, the origin of new ideas. It plays a substantial role in solving problems with extensive and diverse domains, such as best matching—if an analogy is identified between two seemingly unrelated problems, the solution to one of them is indeed a great inspiration for solving the other. For example, an algorithm for parallel machine scheduling can be modified and applied for multi-enterprise network design; a protocol for micro-sensor clustering can be modified and applied for power grid design; a vehicle routing model can be modified and applied for database partitioning; a project management heuristic can be modified and applied for multi-robot team formation.

This chapter presents a taxonomic framework for all the problems in production, manufacturing, and service domains that are reducible to best matching. The framework, called the *PRISM taxonomy of best matching*, was inspired by and developed at the PRISM (Production, Robotics, and Integration Software for Manufacturing and Management) Center of Purdue University. It formalizes best matching problems with respect to 3 + 1 dimensions: D1, sets; D2, conditions; D3, criteria; D+, time or progression (Fig. 2.1). The PRISM taxonomy enables a comprehensive and standardized framework for classifying existing best matching problems and identifying new problem instances in various domains as well as effective problem-solving guidelines, all possible by analogical reasoning. The remainder of this chapter elaborates on the 3 + 1 dimensions of the PRISM

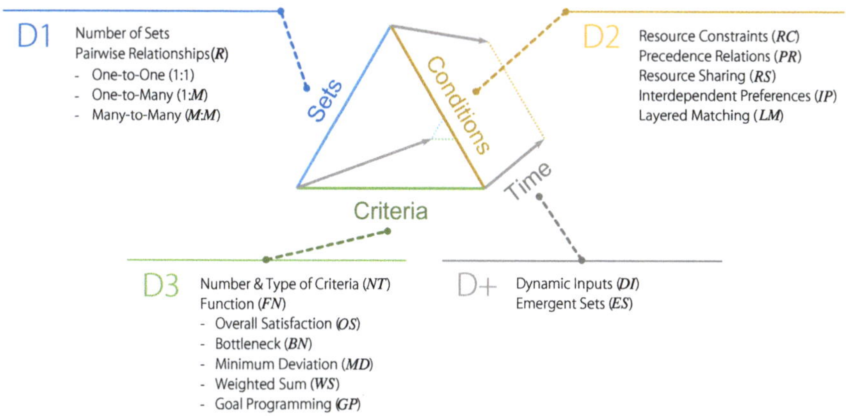

Fig. 2.1 The PRISM taxonomy of best matching (after Moghaddam and Nof 2015a)—a matching process is formally represented by its D1/D2/D3/D+

taxonomy of best matching along with several examples and illustrations. It also explains the prism structure of the four dimensional surfaces and their mutual interactions.

2.1.1 D1: Sets of Individuals

The first dimension of the PRISM taxonomy categorizes the individuals involved in being matched by the matching process into two or more *homogenous* sets; e.g., tasks and processors; students and schools; flights and passengers; items and locations; jobs, machines, and operators.[1] Note that the sets being homogenous does not imply that all the individual set members are the same, but are of the same kind or nature. The first step in formalizing a best matching process is therefore to identify the sets to be matched and their two main characteristics (Fig. 2.2): Number of sets and pairwise relations between individuals (denoted by R). Individuals can be matched to one or more elements of the other set(s)—the pairwise relations can be *one-to-one*, 1:1 (e.g., organs and patients), *many-to-one*, M:1 (e.g., tasks and agents), or *many-to-many*, M:M (e.g., suppliers and customers). In mathematical terms, each entry of R represents the pairwise relations between the elements of two sets in an ascending order of the sum of their indices, and $|R| = N(N-1)/2$. For example, in matching operators, machines, and jobs with exactly one operator assigned to each machine, $R := 1:1:M$. Accordingly, the

[1]By definition, the minimum number of sets in any matching is two sets. (See Problem 2.1) If all items in one or more sets are identical, their matching is trivial.

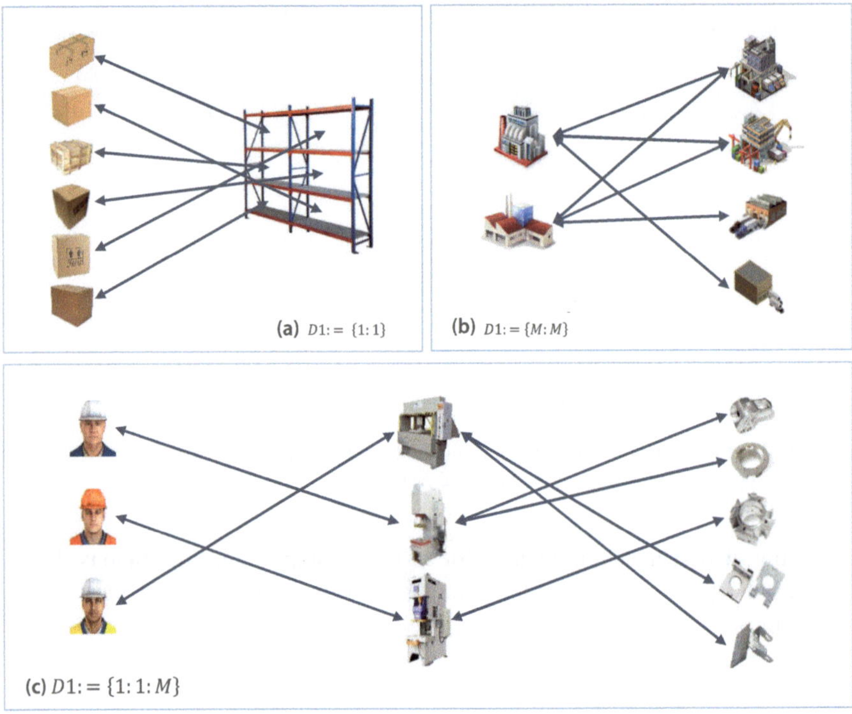

Fig. 2.2 Three matching examples. **a** Boxes and storage locations (e.g., AS/RS)—two sets on one-to-one basis. **b** Manufacturing plants and distribution centers—two sets on one-to-one basis. **c** Operators, machines, and jobs—three sets with one-to-one, one-to-many, and one-to-many relationships, respectively. [↔: matching]

first dimension of a best matching process is formally represented as $D1 := \{R\}$, which entails both the number of sets and their mutual relations.[2]

2.1.2 D2: Matching Conditions

Depending on the context and nature of the system, best matching processes are sometimes "conditional" by certain characteristics, requirements, and/or constraints, which, if disregarded, may lead to erroneous, misleading, inappropriate, or even infeasible matching results. In spite of their significant impact, however, best matching conditions (i.e., D2) have not been established well in the literature compared to the other two main dimensions of best matching (i.e., D1: sets; D3:

[2]Note that *R* also represents the number of sets being matched; i.e., the classification of individuals into two (e.g., tasks and processors; 1:*M*), three (e.g., operators, machines, and jobs; 1:1:*M*), or more sets.

criteria). The second dimension of the PRISM taxonomy of best matching is therefore intended to formalize various conditions[3] that may govern best matching processes. Accordingly, the D2 corresponding to a certain best matching process may involve one, some, or all of the condition types explained below.

(1) *Resource constraints* (*RC*). In matching sets *I* and *J*, the elements of set *I* may require certain amounts of resources from the elements of set *J*, which have limited resources. In Fig. 2.2c, for example, each job (i.e., set *I*) has certain processing time on each machine, while each machine (i.e., set *J*) requires limited production time in each period. The best matching results are therefore influenced by this condition, where the impacts are a function of the resource requirements of the jobs and the available resources of the machines.
(2) *Precedence relations* (*PR*): In matching sets *I* and *J*, the elements of set *I* may have certain precedence relations in their matching to the elements of set *J*. This condition may significantly influence the feasibility and optimality of best matching decisions, if the process also involves *RC*. For example, in assembly line balancing, i.e., matching tasks (set *I*) and workstations (set *J*), since each workstation can accommodate a limited number of tasks (i.e., *RC*), the set of feasible workstations for task $i_1 \in I$ is influenced by the matching of its preceding task $i_2 \in I$ (e.g., if i_2 is assigned to Workstation #2, i_1 cannot be assigned to Workstation #1 at the same time).
(3) *Resource sharing* (*RS*). In matching sets *I* and *J* with *RC* (e.g., *I*: tasks that require resources; *J*: machines with limited resources), the elements of set *J* may not be isolated, but collaborating by laterally sharing their resources with each other. Hence, the problem is to match sets *I* and *J* with respect to their resource requirements and constraints, respectively, given that the elements of set *J* (e.g., machines) are allowed to share their resources among themselves, under certain circumstances and conditions. Consider, as an example, a supply network where the supply enterprises are allowed to collaborate through *demand and capacity sharing* (see, e.g., Kutanoglu and Mahajan 2009; Yoon and Nof 2010; Paterson et al. 2011). In this specific case (Fig. 2.3), the optimal allocation of customers/demand points (i.e., set *I*) to the supply providers (i.e., set *J*)—which is done based on the correspondence between demand and capacity parameters—may be drastically influenced by the lateral collaboration decisions (Moghaddam and Nof 2014). Best matching with *RS* is indeed a special class of best matching with *RC*, where the elements of set *J* are allowed to share resources.
(4) *Interdependent preferences* (*IP*). Matching between sets *I* and *J* is essentially performed based on specific criteria, which represent the mutual *preferences* of the individuals of different sets. The matching preferences are typically fixed input parameters representing the status and characteristics of the system; e.g.,

[3]Note that although all the known and most common matching conditions—both classic and emerging—are covered in this section, D2 may also involve other types of conditions that are not discussed here.

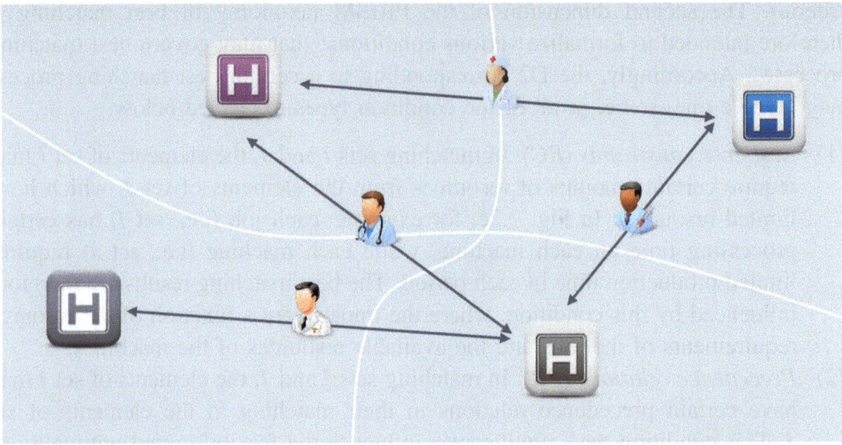

Fig. 2.3 Resource sharing in a collaborative community healthcare network. The "delineation" of the specific region that a community hospital serves, i.e., matching people (set *I*) and hospitals (set *J*) based on geographical location, is noticeably influenced by the ability of the hospitals to mutually share their resources (e.g., doctors) in case of imbalance in the regional capacity and demand. [↔: resource sharing]

cost of production; speed of delivery; quality of service; dimensional fitness. In some cases, however, the preferences are not fixed and independent of environmental and decisional factors (Moghaddam & Nof, 2015a). In two-sided matching between sets *I* and *J*, for instance, the preferences of $i_1 \in I$ for $j_1 \in J$ and $j_2 \in J$ may vary, depending on which other elements of set *I* have already been matched to j_1 and j_2; e.g., matching i_2 to j_1 may lead to increase, decrease, or no change in the preference of i_1 for j_1 (Fig. 2.4). For example,

- The initial preferences of students (i.e., set *I*) for certain course project teams (i.e., set *J*) may be influenced by the affiliation of strong/weak students to those teams;
- The initial preferences of jobs (i.e., set *I*) for different machines (i.e., set *J*) may be influenced by the allocation of other jobs with similar/different setups to those machines;
- The initial preferences of companies (i.e., set *I*) for certain suppliers (i.e., set *J*) may change, depending on the matching of other companies (collaborator or competitor) to those suppliers.

(5) *Layered matching* (*LM*). In certain cases, there are several necessary steps of matching. A typical case: The outcomes of a set of parallel matching processes provide the basis for a subsequent matching process. For example, in an assembly process component parts are matched according to their mutual dimensional fitness, which is considered as the matching criterion. The combination of parts (sets) *I* and *J* produces sub-assemblies *Y*, and the combination of parts *K* and *L* produces sub-assemblies *Z*; then, pairs of *Y* and *Z* are assembled in the next stage to yield the final products (Fig. 2.5).

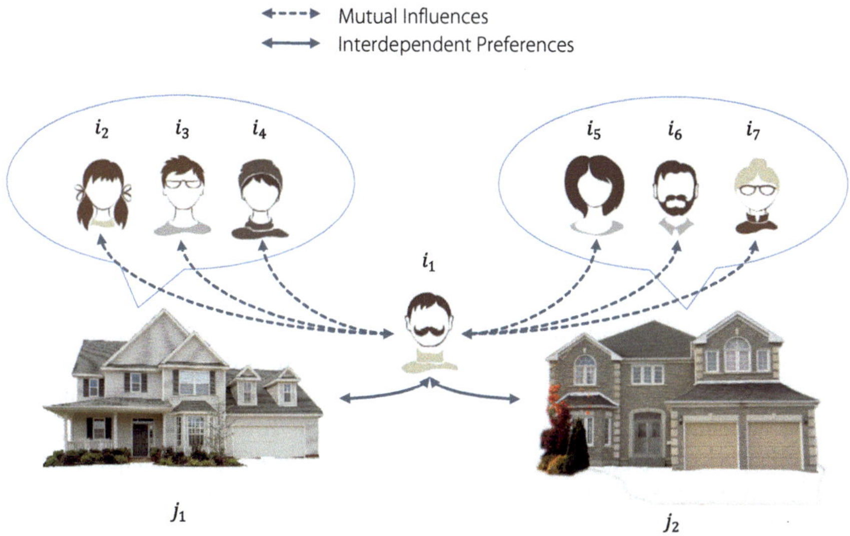

Fig. 2.4 Best matching with interdependent preferences (*IP*). Example: Mr. i_1 has to select from two houses (i.e., j_1 and j_2) and move into it. He has initial preferences for each choice (solid arrows). However, those preferences may be strengthened/weakened, depending on his impression and feelings (dotted arrows) for the current residents already settled in those houses (i.e., $i_2, \ldots, i_7$) (adapted from Moghaddam and Nof 2015a)

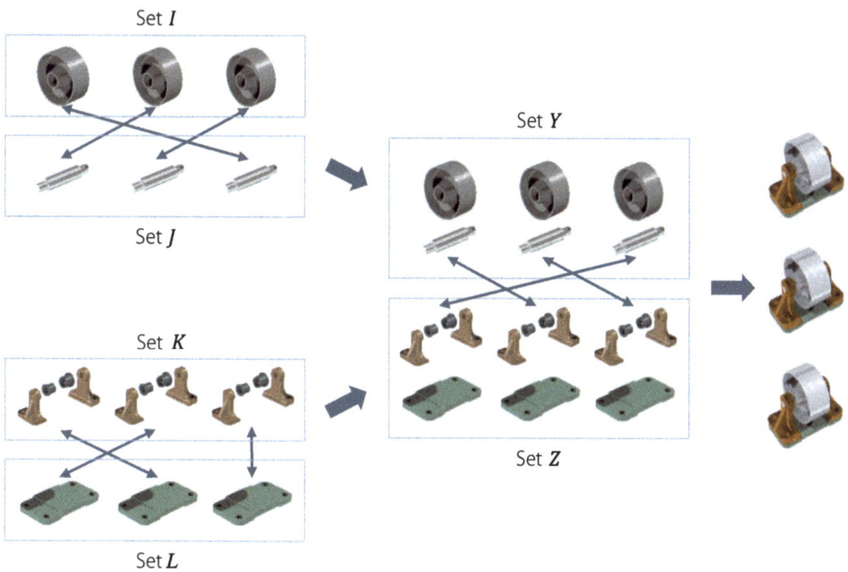

Fig. 2.5 Layered best matching in selective assembly of spin part (courtesy of IronCAD). The elements of sets *Y* and *Z*, resulted from matching sets *I* and *J* and sets *K* and *L*, respectively, in an earlier stage, are matched to produce the final products. The quality of the second stage matching (i.e., *Y* and *Z*) in terms of dimensional fitness is directly influenced by the preceding parallel matching processes (i.e., *I* and *J* & *K* and *L*). [$\leftrightarrow$: matching]

A best matching process may involve one (or more) of the aforementioned conditions. Accordingly, the second dimension of the PRISM taxonomy of best matching may include *RC*, *PR*, *RS*, *IP*, and/or *LM*. For the best matching processes shown in Figs. 2.3, 2.4 and 2.5, the second dimension of the PRISM taxonomy is represented as D2 := $\{RC, RS\}$, D2 := $\{RC, IP\}$, and D2 := $\{LM\}$, respectively.

2.1.3 D3: Matching Criteria

The *best match* is naturally distinguished from other potential matches based on a set of criteria (one or more). The goal is then to enhance the performance of the system with respect to the given criterion/criteria by optimally matching the distributed individuals. In the range of possible matches, between the mismatch and the optimal match, there are possibilities of compromise match, adequate match, and best match. Optimal match can be found only if all matched components are perfect. In practice, however, best match means the best that can be accomplished, including the case of optimal match when it is feasible. Benefits of best matching, in terms of the resulting best match, include advantages gained in the quality, effectiveness and value of design, decisions, services, products; and in the efficiency and timeliness of plans, procedures, processes, activities and responses. With respect to the matching process itself, the best matching process can be measured by its timeliness, reliability, cost-effectiveness, and so on.

The definition of "best," both for the matching process and for the resulting best match depends on the matching objectives (criteria). Various such objectives are defined and discussed throughout this book. For instance, recall the example of matching bolts and nuts (see Fig. 1.1 in Chap. 1). Suppose we want to match a set of bolts and a set of washers, such that the dimensional gap between each pair when assembled is minimal. We assume all the parts are within design tolerance, but a tighter fit would result in better quality (see Velásquez and Nof 2008, 2009). Best matching criteria are diverse and depend on the context and application domain—from traditional cost, time, efficiency, and productivity factors to emerging e-Criteria (Nof 2007) such as:

(1) *Integrability*. Ability to integrate data from a number of distributed entities and increase their relative usefulness.
(2) *Connectivity*. Type, level, and quality of ICT-supported connections between distributed operating systems, and application and network layers.
(3) *Agility*. Ability of a system, at individuals or network level, to respond and adapt to changes in real-time.
(4) *Scalability*. Ability of a process, system, or network to handle increasing amount of tasks and adapt to growth.
(5) *Reachability*. Effectiveness of interconnections and interactions between individuals in a distributed network.

(6) *Viability*. Ratio of the cost of operating/sustaining distributed individuals to the rewards gained from their respective service.
(7) *Autonomy*. Level of delegation of authority, task assignments, and decentralization in distributed networks.
(8) *Dependability*. Probability of a task to be successfully executed—including system *availability*; *reliability*; *sustainability*; *integrity*; and *maintainability*.
(9) *Resilience*. Ability to survive the unforeseen circumstances, risks, disruptions, and high impact events (*a.k.a.*, *transformability* and *adjustability*).

Although the definitions of some criteria may be completely different, their implications are the same—*desirable* criteria (e.g., agility) must be increased/strengthened while *undesirable* criteria (e.g., cost) must be decreased. Hence, without loss of generality, all matching criteria can be formalized as *preferences* of individuals for each other. Specifically, the quality of matching individuals $i \in I$ and $j \in J$ is represented by their mutual preference as

$$P_{ij} = w_i.p_{i \to j} + w_j.p_{j \to i}, \quad \forall i \in I, j \in J. \tag{2.1}$$

where $p_{i \to j}$ and $p_{j \to i}$ denote the *normalized* preferences of $i \in I$ for $j \in J$ and $j \in J$ for $i \in I$, respectively. Coefficients w_i and w_j, $w_i + w_j = 1$, denote the relative *weights* of individuals i and j in defining the mutual preferences.

The third dimension of the PRISM taxonomy of best matching formalizes the matching criteria based on the mutual preference scores. Specifically, D3 provides information on the number and type (desirable: "+"; undesirable: "−") of criteria (*NT*) along with the formulation of the respective objective function (*FN*):

- Single criterion

  (1) *Overall Satisfaction* (*OS*). This classic function maximizes the overall satisfaction of preference scores, e.g., minimizing the total cost of production (D3 := $\{-, OS\}$); maximizing the overall service level (D3 := $\{+, OS\}$); minimizing the total transportation time (D3 := $\{-, OS\}$).
  (2) *Bottleneck* (*BN*). This function maximizes the minimum satisfaction degree of individuals. Application examples of bottleneck assignment include (Ravindran and Ramaswami 1977):

     a. Matching printing jobs and press machines in order to minimize the makespan (D3 := $\{-, BN\}$).
     b. Transportation of perishable goods from warehouses to markets with minimum spoilage (D3 := $\{-, BN\}$).
     c. Shipment of military supplies from warehouses to command posts in case of emergency (D3 := $\{-, BN\}$).

  (3) *Minimum Deviation* (*MD*). This function minimizes the gap between the maximum and minimum satisfaction degrees (Martello et al. 1984) or the

average and minimum satisfaction degrees (Gupta and Punnen 1988). Application examples include (Duin and Volgenant 1991):

a. Cooperation between competitors on the construction of a communication network with the objective of minimizing the gap between the maximum and minimum construction and future maintenance cost (D3 := $\{-, MD\}$).
b. Allocation of patients with different degrees of condition severity to different test groups such that the gap between the maximum and minimum severity conditions is minimized (D3 := $\{-, MD\}$).
c. Assignment of tasks to machines in a project with multiple independent phases, where busy machines cannot process the tasks of other phases until the current process is finished, and the objective is to minimize the machines idle times (D3 := $\{-, MD\}$).

- Multiple criteria

(1) *Weighted Sum* (*WS*). This function combines the normalized values of all criteria into one—with respect to their type (i.e., desirable/undesirable)—and optimizes all criteria simultaneously via the unified function. An application example is the assignment of referees to football matches in an Italian championship (Scarelli and Narula 2002) based on multiple criteria (four in this example) such as reliability, evaluation of fitness, international prestige, and refereed matches/number of years worked (D3 := $\{++++, WS\}$).
(2) *Goal programming* (*GP*). This function prioritizes the criteria and considers them in sequence, based on certain target values and bounds. An application example is reallocation of remedial education teachers from the schools at which they taught in the morning to the schools at which they are to teach during the afternoon (Lee and Schniederjans 1983). The matching criteria for this matching instance were (a) the cost of travel between the schools, (b) the mutual preferences of teachers and schools, and (c) the recommendations of the teachers' supervisors, where (a) $\gg$ (b) $\gg$ (c) (D3 := $\{-++, GP\}$).

The objective of a best matching process can be formalized through one of the aforementioned (or similar[4]) functions. Note that based on the definition of collaboration and competition in Chap. 1, these functions are essentially useful for collaborative best matching where all individuals pursue a set of global objectives

[4]Note that although all common criteria functions are included in the third dimension of the PRISM taxonomy of best matching, the criteria functions may be formulated in other formats that are not discussed here.

that can be formulated in a single unified function. The properties and differences of competitive and collaborative best matching and their respective mechanisms are described in detail in Chaps. 4 and 6 of the book.

2.1.4 D+: Time or Progression

All three primary dimensions of a best matching process may be influenced by *time*, certain progression of time, steps, phases, frames, as an additional dimension. The dynamic and emergent behaviors of systems and their environments may evolve and change the characteristics of the sets (e.g., number of sets and elements; mutual relationships), conditions (e.g., dynamic interdependencies between preferences), and/or criteria (e.g., changes in definitions/values). Dynamic updates in the sets of subassembly parts with various dimensional tolerances (Velásquez and Nof 2008) or dynamic allocation of customers to enterprises based on their association/dissociation to/from a Collaborative Networked Organization (CNO) (Moghaddam and Nof 2015b) are among several examples of dynamic best matching processes. This additional dimension (D+) requires effective real-time optimization and control mechanisms to uphold the quality of matching over time. D+ characterizes the time-based (progression-based) behavior of best matching processes with respect to two main properties:

- *Dynamic Inputs* (*DI*). Some best matching processes may involve variations in the input parameters in real-time. For example, clusters formation in wireless sensor network is regarded as a best matching process with dynamic inputs (*DI*), because the energy levels of sensors—as one of the main criteria for matching sensors to peers/clusters—vary over time.
- *Emergent Sets* (*ES*). The number of individuals in one or more sets may vary in real-time in certain best matching processes. An example for this case is machine scheduling where the process dynamically undergoes changes in the number of jobs (due to, e.g., variations in demand; introduction of new products; saturation of old products) and/or machines (due to, e.g., breakdowns; maintenance; orders from multiple sources).

If a matching process involves neither *DI* nor *ES*, D+ is left blank in the formal representation of the PRISM taxonomy of best matching. D+ is motivated and addressed by two emerging design principles of CCT, the Collaborative Control Theory (Nof et al. 2015), which will be further elaborated in Part II of the book:

- *Association/dissociation*. In a CNO (e.g., networks of sensors; agents; robots; enterprises), this principle guides the optimality of decisions in terms of (a) the time/duration to collaborate and participate in a CNO; (b) collaborative decisions and respective outcomes; (c) the cost of participation; (d) the reasoning behind remaining in or joining a CNO; (e) the performance achieved through association/dissociation decisions. The association/dissociation principle is relevant to *DI* in the D+ dimension of the PRISM taxonomy of best matching.

- *Emergent lines of collaboration.* This principle guides (a) the quality of collaborative decision-making in complex environments; (b) cost-efficient provision of knowledge and information; (c) development of emergent networks. The principle of emergent lines of collaboration is relevant to *ES* in the D + dimension of the PRISM taxonomy of best matching.

2.1.5 The Prismatic Structure of the PRISM Taxonomy

From the definitions and explanations of the PRISM taxonomy in the previous sections it becomes clear that the three dimensions, D1, D2, and D3 interact mutually during the matching process. Furthermore, along the time or progression dimension, D+, they extend, vary and evolve to a prismatic volumetric structure (Fig. 2.1). The length of the prism represents the length of the matching process. A matching process may be short or long; e.g., completing a scheduling computation in a few milliseconds; completing a jigsaw puzzle in a few minutes; matching team members to their assignments in a few hours or days; competing in a tournament over a season; clashing and matching forces through battles over hundreds of years; exploring space to identify and match planets over many decades; and searching and matching over a lifetime to find again the unforgettable smell and flavor of your favorite childhood dish. A matching process may be completed once, repeated several times, or repeated continuously. It may also remain incomplete. These completion and repetition cases can also be illustrated by the examples in the next sections.

2.2 Four Examples of the PRISM Taxonomy Application

The logic and 3 + 1 dimensions of the PRISM taxonomy of best matching were introduced in the previous section, and several examples were presented for describing the attributes, properties and mechanisms of each dimension. In this section, four examples are selected for detailed and comprehensive illustration of the PRISM taxonomy and D1/D2/D3/D+ notation.

2.2.1 Balancing Collaborative Assembly Lines $(M : 1/RC, PR, RS/ - -, WS)$

Assembly line balancing is a process of assigning a set of tasks to a set of interconnected workstations such that their workload is balanced with respect to the required production throughput. Each task has certain duration and precedence relations with other tasks, cannot be subdivided, and must be processed at a specific workstation. The objective of an assembly line balancing problem is typically to minimize the number of workstations for a given cycle time (Type-I balancing) or minimize the cycle time for a given number of workstations (Type-II balancing). In practice, assembly lines are usually not balanceable, i.e., the overall workload is not equally distributed between all active workstations due to various technical and technological constraints. The consequences of such imbalances are underutilization of resources, bottlenecks, and low efficiency and flexibility of assembly lines. An emerging solution to shorten this gap is *collaboration* among workstations (Moghaddam and Nof 2015c, d), where partially loaded workstations share their idle *tools* (e.g., shareable equipment/workforce) with overloaded workstations during each cycle (Fig. 2.6). The result will then be less idle times and shorter cycles at partially loaded and overloaded workstations, respectively.

This matching problem can be formalized by the PRISM taxonomy of best matching as follows (and as shown above):

- D1 := $\{M : 1\}$
  - Matching a set of tasks (*I*) and a set of workstations (*J*). Tasks cannot be subdivided and must be processed at one workstation. Each workstation can process multiple tasks.
- D2 := $\{RC, PR, RS\}$
  - *RC*: Workstations have limited capacities (cycle time) for processing tasks.
  - *PR*: Tasks have precedence relations for processing.
  - *RS*: Workstations are allowed to share their tools with each other.
- D3 := $\{--, WS\}$
  - Two objectives with different weights are involved in the balancing decisions: (a) Minimize the cycle time; (b) Minimize the number of workstations.
- D+ := None.

2.2.2 *Part Pairing for Concurrent Loading-Machining* $(1 : 1//-, OS)$

Milling is a machining process that utilizes rotary cutters for removing materials from a work-piece. Like other machining applications, all milling processes—in spite of their diversity—begin with loading a work-piece on the machine and possibly certain setting up. During the loading process, the machine is idle, which undesirably imposes considerable costs and inefficiencies in the system. A potentially effective solution to this problem is *dual-pallet machining*—a

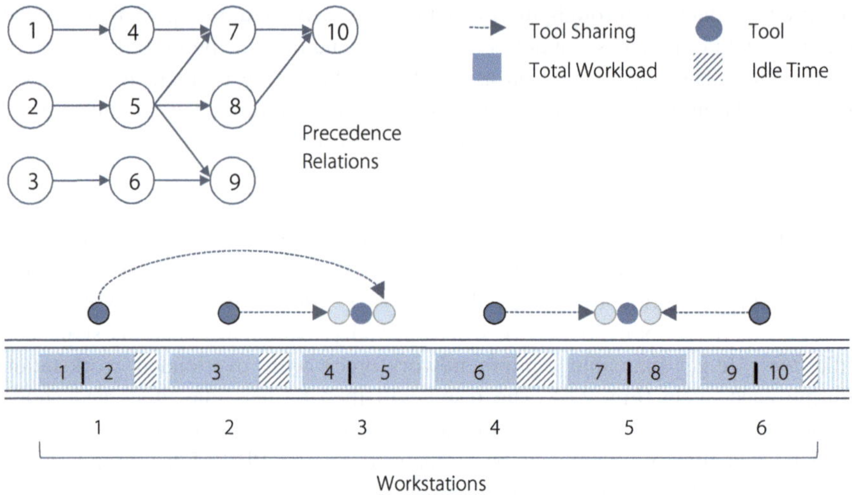

Fig. 2.6 Balancing assembly lines with collaborative workstations (after Moghaddam and Nof 2015c). The precedence relationships along with cycle time upper bounds (i.e., resource constraints) limit the set of feasible workstations for some tasks; e.g., Task 9 cannot be matched to Workstations 1–3, because its immediate predecessors have been matched to Workstations 3 and 4. The workstations are allowed to laterally collaborate through sharing their tools in order to enhance balanceability and resource utilization

configuration that enables concurrent loading and machining of work-pieces, thus minimizing idle time and inefficiencies. The work-pieces on the loading and machining pairs of dual-pallets are not necessarily identical, which makes it difficult to fully materialize the potentials of this configuration. That is, if the loading and processing times of the parts on each one of a dual-pallet pair are significantly different, the process still suffers considerable idle times. For example, if the parts on Pallet #1 require processing time of 4 min, and the parts on Pallet #2 require loading time of 10 min, the machine will be idle for 6 min after processing a part from Pallet #1 (Nof et al. 1979).

Consider a milling process with dual-pallet machining configuration (Fig. 2.7a), where different types of work-pieces are assigned to similar pallets (the same size) and have specific loading and machining times. Due to limitations in the number of fixtures, Pallets #1 and #2 cannot contain similar work-pieces. The total idle time of the milling machine can therefore be minimized by finding the best one-to-one match between set I (i.e., set of work-pieces) and itself, based on loading-machining time gaps, subject to not matching the same element of I (i.e., work-piece) to itself (Fig. 2.7b). This matching problem can be formalized by the PRISM taxonomy of best matching as follows:

- D1 := $\{1 : 1\}$
  - Matching the set of work-pieces (I) to itself. Each work-piece must be matched to one other work-piece for concurrent loading and machining.

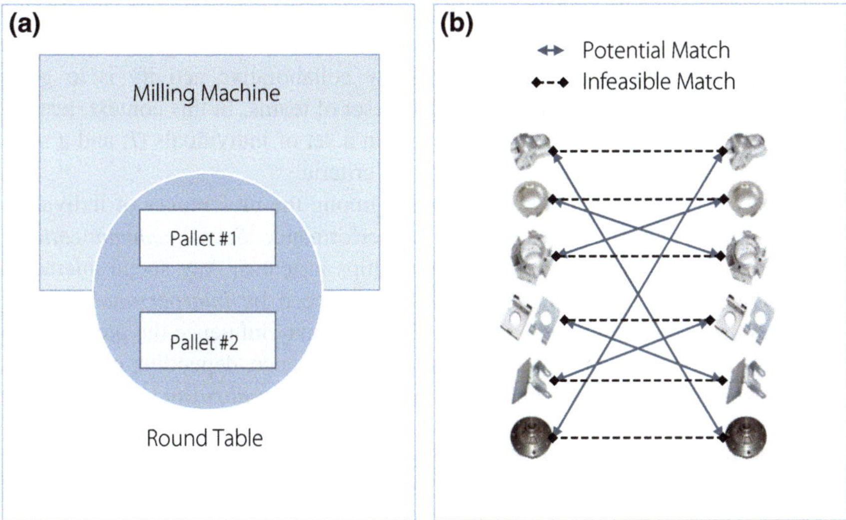

Fig. 2.7 **a** Double-part milling machine configuration (courtesy of Kirby Risk Precision Machining)—Pallets #1 and #2 are under the machining and loading processes, respectively. **b** Examples of potential and infeasible matching. A work-piece cannot be matched to itself. The matching is "symmetric"; i.e., in matching I_L (*left-side* set) and I_R (*right-side* set), matching $A \in I_L$ and $B \in I_R$ implies matching $B \in I_L$ and $A \in I_R$ as well

- D2 := None[5]
- D3 := $\{-, OS\}$
  - The objective is to minimize the gap between loading time of the work-pieces in Pallet #1 and processing time of the work-pieces in Pallet #2, and vice versa. In order not to match a work-piece to itself, the corresponding gap parameters must be replaced by sufficiently large numbers in the objective function.
- D+ := None.

2.2.3 *Dynamic Teaming with Interdependent Preferences* $(M: 1/RC, IP/+, OS/DI, ES)$

A team is a collaborative network of individuals (e.g., genomes; social insects; humans; sensors; machines; projects; factories) that interact through sharing information, resources, and responsibilities to accomplish a set of common goals

[5]The matching process does not involve any specific conditions, except for not matching a work-piece to itself, which can be handled by slight modifications in the objective function.

(i.e., team members are incented to collaborate). According to Tuckman's forming-storming-norming-performing model of group development (Tuckman 1965), the very first step in conducting any collaborative activity is to group homogenous/heterogeneous individuals into a set of teams. In this context, teaming refers to a best matching process performed on a set of individuals (*I*) and a set of teams (*J*), considering certain conditions and criteria.

In a network of teams, *interdependencies* among the preferences of individuals can have significant impacts on quality and performance. *Social communication* is essential for creating and sustaining relationships in teams. Any social interaction and its aggregated effectiveness are indeed influenced by *interpersonal relationships*, *relational messages*, and *emotion*. Emotions may influence the preference of individuals for a team and subsequently their performance, depending on *who their teammates are*. Such mutual influences are classified as *altruism*, *envy*, and *neutrality*—if Person A is altruistic/envious/neutral about Person B, and Person B is already a member of Team Z, the preference of Person A for Team Z is increased/decreases/not changed (Moghaddam and Nof 2015a). To better understand this concept, recall the example of Mr. i_1 who was looking for a new house (Fig. 2.4), where his initial preferences for each house ($\equiv$ team) were influenced by the people ($\equiv$ team members) already living at each house.

Teams are typically formed in a way that the overall performance, harmony, and/or satisfaction of individuals are optimized. This process, however, may evolve over time; i.e., involve changes in the individuals' preferences, performance, and mutual influences, and/or the number of individuals/teams. Examples for the latter case, association/dissociation behavior, include arrival/departure of a job to/from a production cell; allocation of new interns/students to hospitals/schools; temporary or permanent loss of a team member or leader; opening/closing a line in a service system such as a bank; establishing a new school/hospital (Fig. 2.8). Given all these cases, the team formation problem with interdependent preferences can be formalized by the PRISM taxonomy of best matching as follows:

- D1 := $\{M : 1\}$
  - Matching a set of individuals (*I*) and a set of teams (*J*). Each individual must be a member of exactly one team, while each team may be composed of multiple individuals.
- D2 := $\{RC, IP\}$
  - *RC*: Each team can accommodate up to a certain number of individuals.
  - *IP*: There are interdependencies among the preferences of individuals for teams.
- D3 := $\{+, OS\}$
  - The objective is to maximize the overall performance of the teams, or the overall satisfaction of the individuals.
- D+ := $\{DI, ES\}$

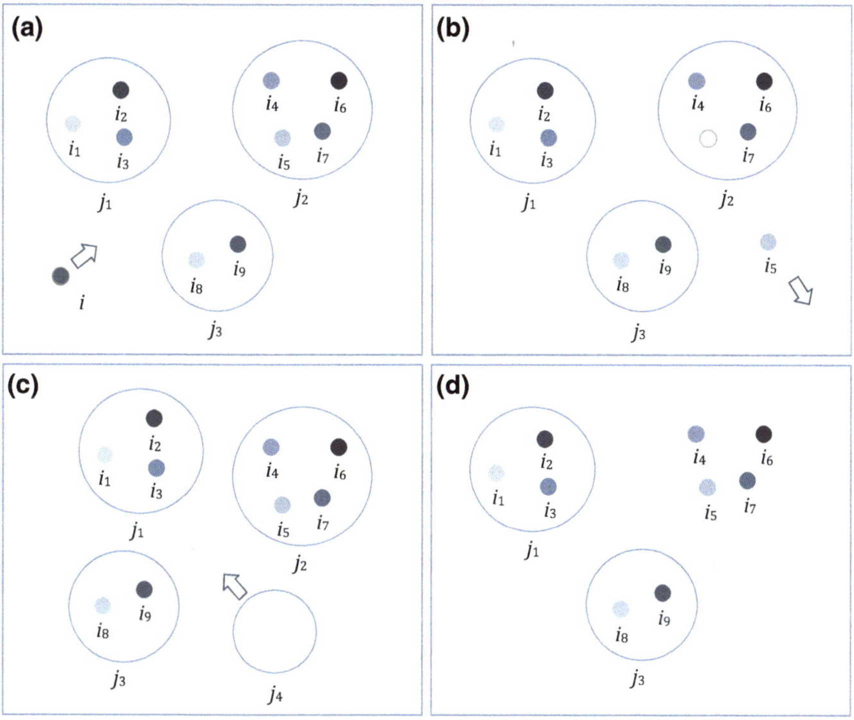

Fig. 2.8 Evolution of a network of teams; e.g., dynamic alliances of enterprises. **a** Association or a new enterprise to the network of alliances ($i_{10} \in I$). **b** Dissociation of an existing enterprise from the network of alliances ($i_5 \in I$). **c** Formation of an alliance ($j_4 \in J$). **d** Dissolution of an alliance ($j_2 \in J$) (after Moghaddam and Nof 2015a). [⇨ : Association/Dissociation]

- *DI*: The individuals' preferences for teams and their mutual influences on each other, the teams' preferences for individuals and their capacity limits, and other system's characteristics may vary over time.
- *ES*: New individuals/teams may *associate* with the system and old individuals/teams may dissociate from it (Fig. 2.8).

2.2.4 Location-Allocation Decisions in CNO $(1 : M : M/RC, PR, RS/+-, WS)$

CNO (Collaborative Networked Organizations) are organizations—systems of resources deployed for processing and synthesizing a set of tasks—of autonomous, distributed, and comprising heterogeneous individuals that collaborate through sharing information, resources, and responsibilities to achieve common and compatible goals (Camarinha-Matos et al. 2009). The definition of CNO stands for a wide variety of systems. For example, (see Moghaddam and Nof 2015e)

- A procurement system with a network of order, product, and resource agents.
- A community healthcare network with a set of hospitals, equipment, personnel, and doctors.
- An intelligent warehouse system with a network of sensors/RFID tags, readers, and antennas.
- A virtual factory with a set of decoupled software tools, shop floor devices, communication and computational elements, middleware, knowledge bases, and servers.
- A manufacturing network with a set of suppliers, distributers, and manufacturing plants.

To address the emerging needs of CNO for higher scalability, resilience, and flexibility (Putnik et al. 2013), interaction and collaboration among distributed *organizations*, *resources*, and *tasks* must be carefully designed and optimized. With the rising dominance of e-Work and e-Activities in manufacturing, business, and service (Nof et al. 2015), organizations are rapidly transforming into cyber-physical networks of resources, shared through cloud-based information and communication channels (Xu 2012) for concurrent processing of remote tasks (e.g., cyber control of tele-robots; virtual manufacturing; remote modeling/simulation/problem-solving). These emerging features demand a new definition of the classic *location-allocation problem* (Fig. 2.9):

In a CNO with a set of organizations, resources, and tasks, what are the best organization to 'locate' each resource and the best set of resources to 'allocate' each task, given that organizations can collaborate by sharing their resources and tasks?

The above problem is known as *Collaborative Location-Allocation Problem* (CLAP) (Moghaddam and Nof 2015b). The CLAP is indeed a problem of 3-way

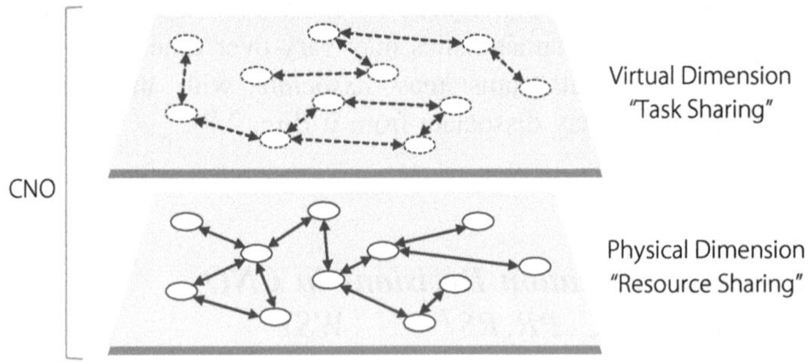

Fig. 2.9 Best matching enhances the performance of CNO at both physical and virtual/cyber levels by optimal location of resources and allocation of tasks over time (after Moghaddam and Nof 2015b). [O: organization]

matching of organizations, resources, and tasks. The PRISM taxonomy of best matching formalizes the CLAP as follows:

- D1 := $\{1 : M : M\}$
  - Matching a set of organizations (*I*), a set of resources (*J*), and a set of tasks (*K*). Multiple resources can be assigned to one organization, while each individual resource can be located in one organization. Each task must be assigned to exactly one organization, while an organization can process multiple tasks. Each resource can be used for processing multiple tasks and each task can be processed by multiple resources.
- D2 := $\{RC, PR, RS\}$
  - *RC*: Each organization can accommodate a limited amount of resources and each resource can process a limited amount of tasks.
  - *PR*: Tasks may have precedence relations for processing.
  - *RS*: Organizations are allowed to share their shareable resources and tasks with each other, given potential capacity shortage/surplus over time due to dynamic variations in demand and capacity.
- D3 := $\{+-, WS\}$
  - Different objectives can be considered in the CLAP; e.g., maximizing task fulfillment rate and minimizing collaboration cost, each with specific weights.
- D+ := None.

2.3 Summary

The PRISM taxonomy of best matching presented in this chapter provides an effective foundation for synthesis and formal representation of best matching processes with respect to 3 + 1 dimensions. The motivation lies in its three functions (Fig. 2.10):

(1) *Identification*. With concrete decomposition of a best matching process with respect to the characteristics of the sets of individuals, potential matching conditions, criteria, and real-time behavior, the PRISM taxonomy provides a powerful tool for scrutinizing various processes and systems in different manufacturing and service domains, and characterizes them as different classes of best matching processes.

(2) *Specification*. The PRISM taxonomy enables hierarchical representation of best matching processes in various domains by defining and quantifying four dimensions and their sub-categories—it generates standardizes best matching processes, empowers problem-solving capabilities, and helps with identification of new processes.

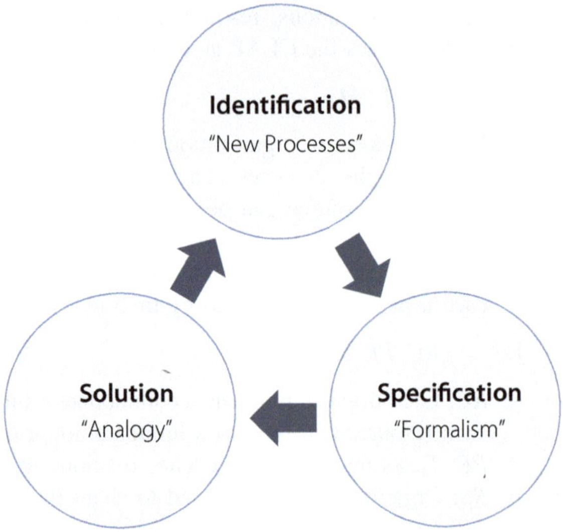

Fig. 2.10 Three functions of the PRISM taxonomy of best matching

(3) *Solution*. The PRISM taxonomy triggers analogical reasoning—as the origin of new ideas and human creativity—by enabling analogical comparisons between best matching processes from completely unrelated domains. This, in turn, leads to inspirations from the already existing solutions to analogous problems for solving unsolved problems.

The taxonomy introduced in this chapter is the basis for the rest of discussions throughout this book. Chap. 3 presents detailed mathematical formulations of various classes of best matching processes, based on the given classification by the PRISM taxonomy. Chaps. 4–6 provide solutions to different classes, and elaborates further on the impact of analogical reasoning on problem solving in this domain.

Problems

2.1. Discuss several reasons why the authors chose the prism shape for The PRISM taxonomy of best matching. Include also the optical perspective: It is represented as a "prismatic taxonomy" (Fig. 2.1) by a triangular geometric shape of a prism with rectangular sides; such optical prisms are used for light reflection/refraction.

2.2. Compare and contrast problems that can be solved by a single matching solution process, and those that require repeated matching solutions.

2.3. Explain the factors that influence the duration of a single matching process. Can this duration be shortened? Explain how.

2.4. Consider the robustness of the match, and explain how you can increase this robustness.

2.5. Consider the matching taxonomy Sect. 2.2.1. How would the taxonomic representation change in each of the following cases (each case is independent of the others)?

(a) There is no tool sharing.
(b) Tool sharing is allowed only between pairs of contiguous workstations; i.e., 1 and 2; 2 and 3; 3 and 4, etc.
(c) The line can be designed with only two workstations.
(d) There are no precedence relations among tasks.
(e) The workstations are self-improving by machine learning.

2.6. Explain the meaning of the following taxonomic representations in the context of Sect. 2.2.1:

(a) $1 : 1/RC, RS/-, BN/DI$
(b) $M : 1/RC, PR/+-, GP$

2.7. How would the taxonomic representation in Sect. 2.2.2 change in each of the following cases (each case is independent of the others)?

(a) The round table has six pallets for pre-loading of parts (work-pieces).
(b) In addition to matching of work-pieces, they have to be matched to available times based on the maintenance schedule of the milling machine.
(c) The work-pieces have precedence relations with other work-pieces.
(d) Each work-piece has to also be matched with its accompanying tool-kit.
(e) Historical quality inspection results from machining of each work-piece type guide adaptive machining process planning for future work-pieces.

2.8. Explain the meaning of the following taxonomic representations in the context of Example 2.2.2:

(a) $M : 1//+, BN$
(b) $1 : 1//-, OS/ES$

2.9. Consider a warehouse system where various items continuously arrive and have to be assigned to different storage locations. The storage allocation heuristic must follow two main criteria: (1) Frequency of arrival/departure (*a.k.a.*, ABC classification); (2) Mutual “affinity” of items (i.e., the possibility of being ordered together).

(a) Identify the matching processes associated with this system.
(b) Formalize the matching problem(s) identified in (a) using the PRISM taxonomy of best matching.
(c) How many sets of individuals are involved? What are the matching conditions and criteria? Is there any dynamicity (time/progression) involved in the process? Explain your answers.

(d) Which of the four examples of the PRISM taxonomy application (Sect. 2.2) better characterizes this problem? Justify your answer.
(e) Is there any relationship between the notions of product affinity and interdependent preferences? Explain your answer.

2.10. Answer Problem 2.9 assuming that the items must be carried to their locations using a limited number of AGVs (Automated Guided Vehicles).
2.11. Consider the three matching examples shown in Fig. 2.2. Explain how potential interdependencies among preference may influence the outcomes of those matching processes.
2.12. Consider the following cases associated with the example of selective assembly of spin parts shown in Fig. 2.5:

- *Case 1*: Sets *I* and *J* are matched to generate Set *Y*. Sets *K* and *L* are matched to generate Set *Z*. Sets *Y* and *Z* are matched to generate the final products.
- *Case 2*: The elements of Sets *I*, *J*, *K*, and *L* are matched simultaneously to generate the final products.
- *Case 3*: Sets *I* and *L* are matched to generate Set *U*. Sets *K* and *J* are matched to generate Set *W*. Sets *U* and *W* are matched to generate the final products.

Identify the conditions in which the outcomes of the following cases are the same or different: (a) Cases 1 and 2; (b) Cases 1 and 3; (c) Cases 2 and 3; (d) Cases 1, 2, and 3. Explain the potential cause(s) of these similarities/differences.

2.13. Develop several taxonomic representations for the following matching problems. For each, explain your assumptions and specify the meaning of the taxonomic representations.

(a) A simple assembly line balancing problem.
(b) Construction project scheduling.
(c) Assigning emergency response vehicles and crews to locations.
(d) Seating guests at a formal dinner.
(e) Selecting the best name for a new product from a list of proposed names.

2.14. Apply the PRISM taxonomy of best matching to the following problems:

(a) Matching in each of the five fundamental assembly models: chain assembly; base assembly; arithmetic assembly; geometric assembly; and combinatorial assembly (Nof et al. 2013).
(b) Sorting fruits according to quality and size.
(c) Quality inspection in a furniture production facility.
(d) Selective assembly of jet engines.
(e) Testing knowledge of recruits.
(f) Matching experiments during drug research.
(g) Investment portfolio building.

(h) Error classification.
(i) Selecting recipes for a restaurant menu.
(j) Selecting a surgeon for one's surgery.

2.15. Five matching conditions were presented in this chapter, representing the D2 dimension of the PRISM taxonomy of best matching. Identify and explain 1-3 other matching conditions that are presented here by providing practical examples and illustrations.

2.16. Explain if/how the PRISM taxonomy of best matching enhances understanding of matching processes and development of matching solutions through "analogical reasoning".

References

Camarinha-Matos, L. M., Afsarmanesh, H., Galeano, N., & Molina, A. (2009). Collaborative networked organizations—concepts and practice in manufacturing enterprises. *Computers & Industrial Engineering, 57*, 46–60.

Duin, C. W., & Volgenant, A. (1991). Minimum deviation and balanced optimization, A unified approach. *Operations Research Letters, 10*, 43–48.

Duncker, K., (1945). On problem solving. *Psychological Monographs, 58* (Whole No. 270).

Gick, M. L., & Holyoak, K. J. (1980). Analogical problem solving. *Cognitive Psychology, 12*, 306–355.

Gupta, S. K., & Punnen, A. P. (1988). Minimum deviation problems. *Operations Research Letters, 7*, 201–204.

Kutanoglu, E., & Mahajan, M. (2009). An inventory sharing and allocation method for a multi-location service parts logistics network with time-based service levels. *European Journal of Operational Research, 194*, 728–742.

Lee, S. M., & Schniederjans, M. J. (1983). A multicriteria assignment problem, A goal programming approach. *Interfaces, 13*, 75–81.

Martello, S., Pulleyblank, W. R., Toth, P., & de Werra, D. (1984). Balanced optimization problems. *Operations Research Letters, 3*, 275–278.

Moghaddam, M., & Nof, S. Y. (2014). Combined demand and capacity sharing with best matching decisions in enterprise collaboration. *International Journal of Production Economics, 148*, 93–109.

Moghaddam, M., & Nof, S. Y. (2015a). Best matching with interdependent preferences—implications for capacitated cluster formation and evolution. *Decision Support Systems, 79*, 125–137.

Moghaddam, M., & Nof, S. Y. (2015b). *Collaborative location-allocation decisions in networked v-Organizations*. Research memo: PRISM Center, Purdue University.

Moghaddam, M., & Nof, S. Y. (2015c). Real-time administration of tool sharing by best matching to enhance assembly lines balanceability and flexibility. *Mechatronics, 31*, 147–157.

Moghaddam, M., & Nof, S. Y. (2015d). Balanceable assembly lines with dynamic tool sharing and best matching protocols: A collaborative assembly framework. *IIE Transactions, 47*, 1363–1378.

Moghaddam, M., & Nof, S. Y. (2015e). The collaborative factory of the future. *International Journal of Computer Integrated Manufacturing,*. doi:10.1080/0951192X.2015.1066034.

Nof, S. Y., Barash, M. M., & Solberg, J. J. (1979). Operational control of item flow in versatile manufacturing systems. *International Journal of Production Research, 17*, 479–489.

Nof, S. Y. (2007). Collaborative control theory for e-Work, e-Production, and e-Service. *Annual Reviews in Control, 31*, 281–292.

Nof, S. Y., Ceroni, J., Jeong, W., Moghaddam, M. (2015). Revolutionizing collaboration through e-work, e-business, and e-service. *Springer*, ISBN 978-3-662-45777-1.

Nof, S. Y., Wilhelm, W. E., and Warnecke, H-J. (2013). *Industrial Assembly*. Springer.

Paterson, C., Kiesmüller, G., Teunter, R., & Glazebrook, K. (2011). Inventory models with lateral transshipments: A review. *European Journal of Operational Research, 210*, 125–136.

Putnik, G., Sluga, A., ElMaraghy, H., Teti, R., Koren, Y., Tolio, T., et al. (2013). Scalability in manufacturing systems design and operation: State-of-the-art and future developments roadmap. *CIRP Annals - Manufacturing Technology, 62*, 751–774.

Ravindran, A., & Ramaswami, V. (1977). On the bottleneck assignment problem. *Journal of Optimization Theory and Applications, 21*, 451–458.

Scarelli, A., & Narula, S. C. (2002). A multicriteria assignment problem. *Journal of Multi-Criteria Decision Analysis, 11*, 65–74.

Tuckman, B. W. (1965). Developmental sequence in small groups. *Psychological Bulletin, 63*, 384–399.

Yoon, S. Y., & Nof, S. Y. (2010). Demand and capacity sharing decisions and protocols in a collaborative network of enterprises. *Decision Support Systems, 49*, 442–450.

Velásquez, J. D., & Nof, S. Y. (2008). A best-matching protocol for collaborative e-work and e-manufacturing. *International Journal of Computer Integrated Manufacturing, 21*, 943–956.

Velásquez, J. D., & Nof, S. Y. (2009). Best-matching protocols for assembly in e-work networks. *International Journal of Production Economics, 122*, 508–516.

Xu, X. (2012). From cloud computing to cloud manufacturing. *Robotics and Computer-Integrated Manufacturing, 28*, 75–86.

Chapter 3
Mathematical Models of Best Matching

Abstract This chapter presents the mathematical models of best matching in a progressive manner, from simple one-to-one matching to more advanced matching instances, following the 3 + 1 dimensions of the PRISM taxonomy; i.e., matching processes with different set characteristics, conditions, and criteria, in static or dynamic environments. The purpose is to demonstrate the mathematical formulation procedures, and the evolution of best matching models as the characteristics of the 3 + 1 dimensions of matching processes change and become more complex.

3.1 Why Mathematical Modeling for Best Matching?

A model is "a system or thing used as an example to follow or imitate" (Oxford Dictionaries), and is referred to as a *mathematical model* when presented using mathematical concepts and language. Mathematical models describe the mechanics and behaviors of analogous systems at relatively abstract levels—they support *analogical reasoning*. Specifically, mathematical models help system engineers and designers to:

- Comprehend and explore the meaning of various parameters, variables, and their relationships.
- Develop the basis for mathematical software and programs, and streamline their creation, implementation, and learning processes.
- Estimate, analyze, and compare the quantitative behaviors of the abstracted systems.
- Provide the components for a comprehensive model of a system, as a combination of various conceptual and mathematical models.

The above list provides sufficient reasons for formal representation of an engineering/design problem like best matching in terms of a mathematical model. The next step after formalization of best matching processes through the PRISM taxonomy is therefore to identify its mathematical formulation from the library of best matching models, presented and discussed in this chapter.

M. Moghaddam and S.Y. Nof, *Best Matching Theory & Applications*, Automation, Collaboration, & E-Services 3, DOI 10.1007/978-3-319-46070-3_3

Best matching models are essentially integer or mixed-integer programs, and thus, their development must be in line with the basic principles of mathematical modeling (Fig. 3.1):

(1) The need for the model—what are we looking for?
(2) Objectives—what do we want to know?
(3) Input data—what do we know?
(4) System characteristics—what can we assume?
(5) Leading principles—how do we interpret the model?
(6) Equations, calculations, results—what can we expect from the model?
(7) Consistency with principles and assumptions—are the predictions valid?

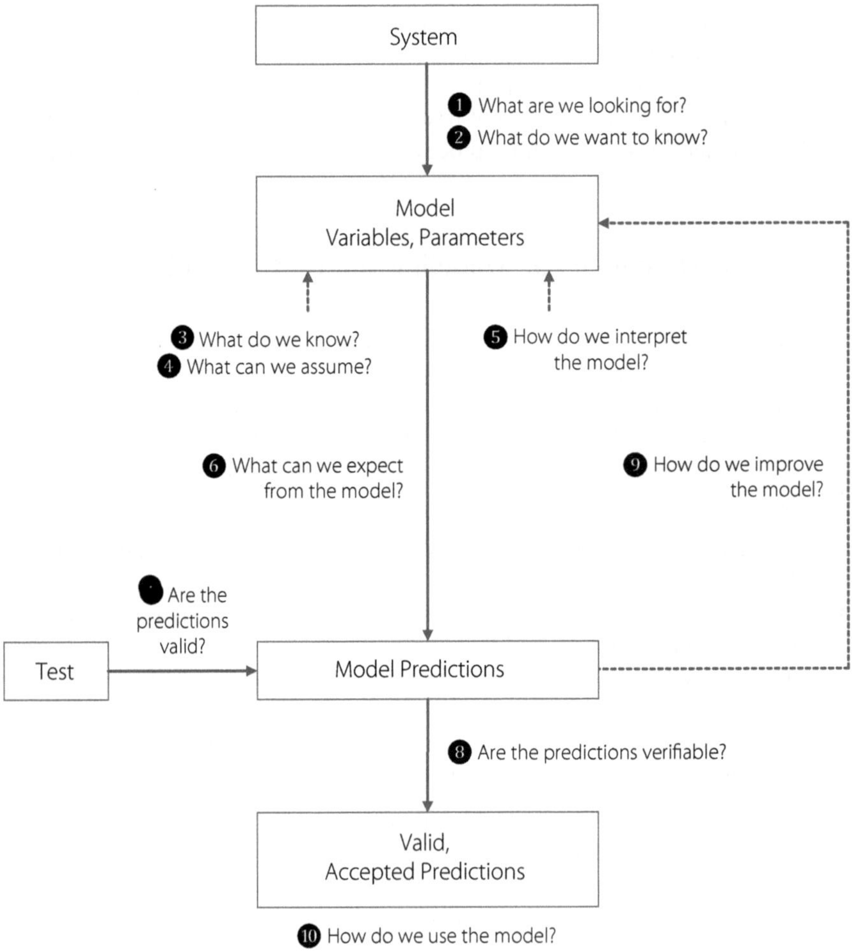

Fig. 3.1 Formal principles and workflow of mathematical modeling (after Cobelli and Carson 2001)

(8) Usefulness in terms of initial purpose—are the predictions verifiable?
(9) Feedback and adjustments in inputs, variables, assumptions—how do we improve the model?
(10) Implementation—how do we use the model?

All best matching models have common properties with reference to the above "checklist". They all search for the best match between two or multiple sets, and represent matching using binary variables χ, where

$$\chi_{ij} = \begin{cases} 1, & \text{if } i \text{ and } j \text{ are matched} \\ 0, & \text{otherwise} \end{cases}, \quad \forall i \in I, j \in J. \tag{3.1}$$

$$\chi_{ijk} = \begin{cases} 1, & \text{if } i, j, \text{ and } k \text{ are matched} \\ 0, & \text{otherwise} \end{cases}, \quad \forall i \in I, j \in J, k \in K. \tag{3.2}$$

and so on. Each best matching model, however, is tailored to optimize a specific set of objectives, and is conditioned by certain requirements and constraints. In addition, the assumptions, the way each model is formulated and interpreted, the verification, validation, and modification procedures may significantly differ from one case to another. The rest of this chapter is dedicated to the formulation of various best matching models following the PRISM taxonomy of best matching (i.e., D1/D2/D3/D+) along with their structural analysis using the principal mathematical modeling checklist and workflow (Fig. 3.1). The goal is to provide a rich yet concise and structured overview of best matching models throughout its many variations developed over more than six decades.

3.2 D1. Sets

The first dimension of the PRISM taxonomy of best matching deals with the characteristics of the sets to be matched—their numbers (N) and pairwise relationships (R). This section presents the mathematical formulations of various combinations of these inputs. In all the models presented in this section, it is assumed that there are no specific conditions, the criterion to be optimized is the overall satisfaction of individuals based on their mutual preferences, and no dynamicity is associated with the matching process in terms of inputs and number of individuals.[1]

[1]These assumptions are merely made for simplicity of mathematical representations, and can be modified/lifted in different cases, as required.

3.2.1 One-to-One Matching

As the most basic instance of best matching problems, one-to-one matching implies matching two (or more) sets in a way that each individual is matched to exactly one element of the other set (e.g., matching bolts and nuts). Considering P_{ij} as the mutual preference of $i \in I$ and $j \in J$, and χ_{ij} as the matching variable shown in Eq. (3.1), the one-to-one matching model is formulated as follows:

$$\begin{aligned} & 1:1//+,OS \\ \text{Maximize} \quad & \sum_{i \in I} \sum_{j \in J} P_{ij}\chi_{ij} \\ \text{Subject to} \quad & \sum_{i \in I} \chi_{ij} = 1, \qquad \forall j \in J, \\ & \sum_{j \in J} \chi_{ij} = 1, \qquad \forall i \in I, \\ & \chi_{ij} \in \{0,1\}, \qquad \forall i \in I, j \in J. \end{aligned} \tag{M 3.1}$$

The objective function of Model (M 3.1) maximizes the overall satisfaction of all individuals in both sets with respect to their mutual preferences. The first two sets of constraints ensure one-to-one relationship between matching sets I and J. The last set of constraints ensures that all matching variables are binary. Considering its unique mathematical structure, Model (M 3.1) is equivalent to a linear program with the same formulation with the last set of constraints replaced with

$$\chi_{ij} \geq 0, \quad \forall i \in I, j \in J.$$

The reason is that in the optimal feasible solution of the linear program, all variables χ_{ij} take either zero or one values.

A necessary feasibility condition for the above formulation of one-to-one matching problem is that the sizes of both sets are equal; i.e., $|I| = |J|$. In some instances, however, this assumption may not be valid/required, and thus, the pairwise relationship constraints may be replaced with the following inequality constraints:

$$\sum_{i \in I} \chi_{ij} \leq 1, \quad \forall j \in J, \tag{3.3}$$

$$\sum_{j \in J} \chi_{ij} \leq 1, \quad \forall i \in I, \tag{3.4}$$

which imply that some individuals will remain unmatched. An extension of this scenario is k-cardinality assignment problem (Dell'Amico and Martello 1997), where only k pairs of individuals are to be matched ($k \leq \min\{|I|, |J|\}$), i.e.,

$$\sum_{i\in I}\sum_{j\in J}\chi_{ij} = k, \quad \forall i \in I, j \in J. \tag{3.5}$$

That is, replacing the relationship constraints in Model (M 3.1) with Constraints (3.3)–(3.5) yields the k-cardinality problem. An application example for this instance of one-to-one matching is in selective assembly, where the k best elements of each set (e.g., bolts and nuts) are selected and matched on one-to-one basis, in order to yield k products with the highest quality. Another example is matching ambulances (set I) to simultaneous emergency calls (set J), where $|I| \neq |J|$ and $k = \min\{|I|, |J|\}$.

3.2.2 *Generalized Matching*

In certain applications, some individuals can be matched to more than one element from the other set(s). Such instances are considered as generalized matching problem, where the pairwise relationships between the matching sets are essentially many-to-one or many-to-many. Examples include assigning multiple tasks to one machine, multiple sensors/robots to one cluster/team, multiple students/interns to one school/factory, multiple suppliers to multiple retailers, and multiple products to multiple consumers. In a generalized matching problem, the total number of matches for each individual is restricted by a certain limit $M_j, j \in J$. The many-to-one matching between sets I and J with the objective of maximizing the overall satisfaction is formulated below. The only difference compared to Model (M 3.1) is in the first set of constraints: The total number of i elements matched to each $j \in J$ can be more than one and must not exceed a certain number. For example, the number of individuals matched to each team is definitely more than one, but also less than a predefined size limit for the team. The one-to-many and many-to-many instances of generalized matching problem can be formulated in a similar manner to Model (M 3.2).

$$\begin{aligned}
& M : 1//+, OS \\
\text{Maximize} \quad & \sum_{i\in I}\sum_{j\in J} P_{ij}\chi_{ij} \\
\text{Subject to} \quad & \sum_{i\in I}\chi_{ij} \leq M_j, && \forall j \in J, \\
& \sum_{j\in J}\chi_{ij} = 1, && \forall i \in I, \\
& \chi_{ij} \in \{0,1\}, && \forall i \in I, j \in J.
\end{aligned} \tag{M 3.2}$$

3.2.3 Multi-Dimensional Matching

All the matching models discussed so far involve two sets, I and J; i.e., $N = 2$. In some matching processes, however, three or more sets need to be matched at the same time; e.g., matching operators, machines, and jobs (see Fig. 2.2c); matching organizations, resources, and tasks (see Fig. 2.9); matching teachers to classes and timeslots (see Daskalai et al. 2004). In this section, basic formulations of three-dimensional matching problems are presented, which can then be further elaborated in more advanced scenarios. Three-dimensional matching problems are usually formulated as *planar* or *axial* models. The planar formulation is presented below.

$$
\begin{array}{lll}
 & 1:1:1//+, OS\,(Planar\,formulation) & \\
\text{Maximize} & \sum\limits_{i\in I}\sum\limits_{i\in J}\sum\limits_{k\in K} P_{ijk}\chi_{ijk} & \\
\text{Subject to} & \sum\limits_{i\in I}\chi_{ijk} = 1, & \forall j \in J, k \in K, \\
 & \sum\limits_{j\in J}\chi_{ijk} = 1, & \forall i \in I, k \in K, \\
 & \sum\limits_{k\in K}\chi_{ijk} = 1, & \forall i \in I, j \in J, \\
 & \chi_{ijk} \in \{0,1\}, & \forall i \in I, j \in J, k \in K.
\end{array}
\tag{M 3.3}
$$

The first three sets of constraints ensure that each individual is matched to exactly one element from each of the other two sets. Similar to Model (M 3.1), a necessary feasibility condition for the above planar formulation is that $|I| = |J| = |K|$. In some instances, however, these equalities may not necessarily hold, and therefore, one or some of the pairwise relationship constraints may be represented as inequalities (similar to Eq. (3.3) and (3.4)). In addition, if the pairwise relationship between two sets is many-to-one or many-to-many, Model (M 3.3) can be generalized in a similar manner to two-dimensional matching problem, as described in Sect. 3.2.2. A classic application example for planar formulation is scheduling meetings between vendors (set I) and customers (set J) over a set of timeslots (set K) (see Gilbert et al. 1987, for details).

The axial formulation of the three-dimensional matching problem is presented next. The difference between the planar and axial formulations is in their representation of pairwise relationship constraints. A classic example for the axial formulation is assignment of a set of jobs (I) to a set of operators (J) and a set of machines (K) (see Gilbert and Hofstra 1988, for details).

$$
\begin{array}{lll}
 & 1:1:1//+, OS\,(Axial\,formulation) & \\
\text{Maximize} & \sum_{i \in I} \sum_{j \in J} \sum_{k \in K} P_{ijk} \chi_{ijk} & \\
\text{Subject to} & \sum_{i \in I} \sum_{j \in J} \chi_{ijk} = 1, & \forall k \in K, \\
 & \sum_{i \in I} \sum_{k \in K} \chi_{ijk} = 1, & \forall j \in J, \\
 & \sum_{j \in J} \sum_{k \in K} \chi_{ijk} = 1, & \forall i \in I, \\
 & \chi_{ijk} \in \{0, 1\}, & \forall i \in I, j \in J, k \in K.
\end{array}
\qquad (\mathrm{M}\,3.4)
$$

3.3 D2. Conditions

The second dimension of the PRISM taxonomy of best matching represents potential conditions that a best matching process may involve; *RC* (resource constraints), *PR* (precedence relationships), *RS* (resource sharing), *IP* (interdependent preferences), LM (layered matching), and more. Details of these conditions together with several examples and illustrations were presented in Chap. 2. This section provides extensions of the mathematical models presented so far, which incorporate these conditions in the formulation. For the sake of simplicity and without loss of generality, all models presented in this section are two-sided, their matching criterion is optimization of the overall satisfaction of individuals based on mutual preferences, and they incorporate no dynamicity in terms of inputs and number of individuals.

3.3.1 Resource-Constrained Matching

In some generalized matching process, the total number of matches for each individual may be restricted by both resource *requirements* and *limits*. Specifically, in many-to-one matching between sets I and J, each element $i \in I$ may require certain amount of resources from each element $j \in J$ (r_{ij}) with limited resources (l_j). For example, in matching a set of tasks (I) and a set of machines (J), each task may have specific processing times on different machines, and the sum of processing times of the tasks assigned to each machine must not exceed its total operating time. The problems with this kind of conditions on resource requirements and limits are referred to as *resource-constrained matching problems.*[2] The many-to-one matching between sets I and J with resource constrains and the objective of maximizing the overall satisfaction is formulated as follows[3]:

---

[2]Note that a resource-constrained matching problem is essentially a generalized matching problem.

[3]The one-to-many and many-to-many matching instances can be formulated in a similar manner.

$$
\begin{array}{lll}
 & M : 1/RC/+, OS & \\
\text{Maximize} & \sum_{i\in I}\sum_{j\in J} P_{ij}\chi_{ij} & \\
\text{Subject to} & \sum_{i\in I} r_{ij}\chi_{ij} \le l_j, & \forall j \in J, \\
 & \sum_{j\in J} \chi_{ij} = 1, & \forall i \in I, \\
 & \chi_{ij} \in \{0, 1\}, & \forall i \in I, j \in J.
\end{array}
\tag{M 3.5}
$$

The *RC* condition extends the formulation of generalized matching problem by generalizing the formulation of resource constraints. That is, Model (M 3.2) is a special case of Model (M 3.5), where all tasks are similar, i.e., $r_{ij} = r_{i'j}, \forall i, i' \in I$. Other variants of the generalized matching problem include problems with (see Öncan 2007, for details)

- *Various levels/states of resource requirements*. The original problem was defined in the context of machine scheduling—allocation of tasks to machines with different levels of efficiency (Glover et al. 1979). With this definition, the problem turns into a three-dimensional matching problem between three sets of tasks (I), machines (J), and efficiency levels (K), where the resource constraints in Model (M 3.5) are reformulated as follows:

$$
\sum_{i\in I}\sum_{k\in K} r_{ijk}\chi_{ijk} \le l_j, \quad \forall j \in J,
\tag{3.6}
$$

where r_{ijk} denotes the resource requirement of $i \in I$ from $j \in J$, when it is operating at the efficiency level of $k \in K$, $\chi_{ijk} = 1$, if $i \in I$ is matched to $j \in J$ at efficiency level of $k \in K$, and $\chi_{ijk} = 0$, otherwise.

- *Multiple resources*. This extension deals with resource-constrained matching between sets I and J, where each element $j \in J$ has different types of resources, and each $i \in I$ requires certain levels of each resource type (Gavish and Pirkul 1991). This variant of resource-constrained matching problem is formulated by replacing the first constraints set of Model (M 3.5) with the following:

$$
\sum_{i\in I}\sum_{k\in K} r_{ijq}\chi_{ij} \le l_{jq}, \quad \forall j \in J, q \in Q,
\tag{3.7}
$$

where r_{ijq} denotes the amount of resource type $q \in Q$ required by $i \in I$ that is matched to $j \in J$, and l_{jq} denotes the level of resource type $q \in Q$ available at $j \in J$.

3.3.2 *Matching with Precedence Relations*

In many-to-one matching between sets I and J, the elements of set I may have certain precedence relations with each other. That is, element $i \in I$ may not be allowed to be matched to element $j \in J$, if its predecessor $i' \in I$ is not matched to $j \in J$ or its preceding elements in set J. A common example of this matching condition in practice is assembly line balancing, where a set of tasks (I) must be matched to a set of workstations (J) with respect to certain precedence relations (see, e.g., Fig. 2.6). This condition can be incorporated in the previous generalized matching formulations by adding the respective precedence relation constraints. For example, a many-to-one resource-constrained matching between sets I and J, with precedence relations between the elements of I, and overall satisfaction degree as the matching criterion, is formulated as follows:

$$\begin{array}{lll} & M: 1/RC, PR/+, OS & \\ \text{Maximize} & \sum_{i\in I}\sum_{j\in J} p_{ij}\chi_{ij} & \\ \text{Subject to} & \sum_{i\in I} r_{ij}\chi_{ij} \leq l_j, & \forall j \in J, \\ & \sum_{j\in J} j\chi_{i'j} \leq \sum_{j\in J} j\chi_{ij}, & \forall i \in I, i' \in Pd_i, \\ & \sum_{j\in J} \chi_{ij} = 1, & \forall i \in I, \\ & \chi_{ij} \in \{0,1\}, & \forall i \in I, j \in J. \end{array} \tag{M 3.6}$$

In Model (M 3.6), the second set of constraints ensures that the precedence relations between the elements of set I are not violated. (Pd_i denotes the set of predecessors of element $i \in I$.) Precedence relations, in general, reduce the sets of feasible matches for some individuals. In some applications, however, such influences may be formulated in different ways. A common example is *positive/negative zoning* (in, e.g., assembly line balancing; scheduling; storage allocation), where if $i \in I$ is matched to $j \in J$, then $i' \in I$ ($i' \neq i$) should/should not be matched to $j' \in J$. The positive and negative zoning constraints are respectively represented as follows (see, e.g., Simaria and Vilarinho, 2004):

$$\chi_{ij} - \chi_{i'j'} = 0, \quad \forall i, i' \in I, i' \neq i, j, j' \in J. \tag{3.8}$$

$$\chi_{ij} + \chi_{i'j'} \leq 1, \quad \forall i, i' \in I, i' \neq i, j, j' \in J. \tag{3.9}$$

3.3.3 Matching with Resource Sharing

In resource-constrained matching, different levels of resource requirements (by the elements of I) and/or limits (associated with the elements of J) may reduce the overall utilization of resources by causing resource slacks and surpluses. That is, the elements of set J typically encounter either excess or shortage of resources. In supply networks, for example, some suppliers may run out of certain products, while their peers hold excessive inventories in their warehouses. Similar scenarios occur in various manufacturing and service systems, where *resource sharing* enhances overall resource utilization and stability of the network (see, e.g., Figs. 2.3, 2.6, and 2.9). In such cases, the overall satisfaction degree can be increased by enabling the elements of I to be matched to their *more preferred* element(s) in J through increasing their available local resources (Moghaddam and Nof 2014). In spite of its inherent advantages, however, resource sharing may be costly, and thus, its respective frequency and quantity may need to be minimized.

Given these considerations, a resource-constrained matching between sets I and J, with resource sharing[4] and the objectives of maximizing the overall satisfaction (including the total cost of resource sharing) is formulated below. In Model (M 3.7), the first term in the objective function maximizes the overall satisfaction degree of individuals with respect to their mutual preferences, while the second term minimizes the total cost of collaboration through resource sharing with normalized cost parameters.[5] $C_{ii'}$ denotes the unit cost of sharing resources between $j \in J$ and $j' \in J$, and $\lambda_{ii'}$ denotes the amount of resources shared by $j \in J$ with $j' \in J$. The first set of constraints therefore ensures that the overall resource requirements of the elements of I from each $j \in J$, plus the amount of resources shared by other elements of J with $j \in J$, minus the amount of resources shared by $j \in J$ with other elements of J must be less than or equal to its resource limit.

$$
\begin{aligned}
& M : 1/RC, RS/+, OS \\
\text{Maximize} \quad & \sum_{i \in I} \sum_{j \in J} P_{ij} \chi_{ij} - \sum_{j \in J} \sum_{j' \in J} C_{jj'} \lambda_{jj'} \\
\text{Subject to} \quad & \sum_{i \in I} r_{ij} \chi_{ij} + \sum_{\substack{j' \in J \\ j' \neq j}} \lambda_{j'j} - \sum_{\substack{j' \in J \\ j' \neq j}} \lambda_{jj'} \leq l_j, && \forall j \in J, \\
& \sum_{j \in J} \chi_{ij} = 1, && \forall i \in I, \\
& \chi_{ij} \in \{0, 1\}, && \forall i \in I, j \in J.
\end{aligned}
\qquad \text{(M 3.7)}
$$

Note that Model (M 3.7) assumes that all resources with different origins (i.e., elements of J) are similar, and there are no losses during the sharing process. In practice, however, this assumption may not be valid. Consider, for instance, the example of assembly line balancing with tool sharing shown in Fig. 2.6. In that case, when an idle tool (i.e., excessive capacity) is shared between two workstations

[4]Note that a matching problem with resource sharing is essentially a resource-constrained matching problem.

[5]The cost parameters must be normalized in order for the first and the second terms of the objective function to be consistent with each other in terms of units (e.g., revenue/benefit; time saved/time added).

during the remaining time until the end of the cycle, the target workstation cannot utilize the tool 100 % during this period, because a portion of it is wasted for movement of the tool between the workstations. This issue can be addressed by simply incorporating the discount factor $\rho_{j'j} \leq 1$ in the model as follows:

$$\sum_{i \in I} r_{ij}\chi_{ij} + \sum_{\substack{j' \in J \\ j' \neq j}} \rho_{j'j}\lambda_{j'j} - \sum_{\substack{j' \in J \\ j' \neq j}} \lambda_{jj'} \leq l_j, \quad \forall j \in J. \tag{3.10}$$

In Eq. (3.10), the discount factor $\rho_{j'j}$ reduces the amount of resources shared by $j' \in J$ with $j \in J$ to a certain extent, depending on their mutual relationships and interactions.

3.3.4 *Matching with Interdependent Preferences*

In all classes of matching processes, the individuals may influence each other's preferences in different ways. In Chap. 2, examples of many-to-one matching with interdependent preferences (*IP*) were discussed (see, e.g., Fig. 2.4), where the elements of set I may influence each other's preferences if they are matched to the same element of set J. Similar scenarios are also likely in one-to-one and many-to-many matching processes, as well as in processes with higher dimensions. This section presents two mathematical formulations of matching with *IP* for one-to-one and many-to-one matching between sets I and J.[6] The former turns into a special case of the classic *quadratic assignment problem* introduced by Koopmans and Beckmann in 1957.

3.3.4.1 One-to-One Matching with *IP*

In matching sets I and J one a one-to-one basis, the mutual preference of $i \in I$ and $j \in J$ for each other may be changed, if $i' \in I$ and $j' \in J$ are matched together. That is, denoting the initial preference of i and j by P_{ij}, their *IP* (interdependent preference) can be formulated as follows:

$$\widehat{P}_{ij} = P_{ij}\left(1 + \sum_{i' \in I}\sum_{j' \in J} \alpha_{iji'j'}\chi_{i'j'}\right), \quad \forall i \in I, j \in J. \tag{3.11}$$

[6]Other cases can be formulated in a similar manner to one-to-one and many-to-one matching processes.

In the *IP* formula, parameter α represents the impact of matching i' and j' (i.e., $\chi_{i'j'} = 1$) on the mutual preference of i and j. Specifically, (Moghaddam and Nof 2015)

- If i and j are altruistic about i' and j', $\alpha_{iji'j'} > 0$.
- If i and j are envious about i' and j', $\alpha_{iji'j'} < 0$.
- If i and j are neutral[7] about i' and j', $\alpha_{iji'j'} = 0$.

The notions of *altruism*, *envy*, and *neutralism* are borrowed from the terminology of utility theory, which are used to indicate the dependencies of the individuals' preferences on the consumption or well-being of other individuals in their neighborhood (*a.k.a.*, peer influence, neighborhood effect, bandwagon effect, conformity; Yang and Allenby 2003). Given the above definition, the one-to-one matching with *IP* is formulated below. In Model (M 3.8), the objective function is obtained by simply replacing the mutual preference values in the objective function of Model (M 3.1) (i.e., one-to-one matching without *IP*) with the mutual *IP* presented in Eq. (3.11). The nature and mathematical formulation of the one-to-one matching process with *IP* are similar to those of the *quadratic assignment problem*, defined as the problem of locating a number of facilities (set *I*) in a number of locations (set *J*), with specific *distance* between each pair of locations and specific *flow* between each pair of facilities. The objective is to find the best match between facilities and locations such that the sum of the distances weighted by the respective flows between the facilities is minimized. Naturally, the cost/benefit of locating facility i at location j (i.e., mutual preference of i and j) depends on the location of other facilities. Let $f_{ii'}$ and $d_{jj'}$ respectively denote the flow between facilities i and i', and the distance between locations j and j'. The original formulation of the quadratic assignment problem is as follows, which has a similar structure to Model (M 3.8).

$$
\begin{array}{llll}
 & 1:1/IP/+,OS & & \\
\text{Maximize} & \sum_{i\in I}\sum_{j\in J} P_{ij}\chi_{ij} + \sum_{i\in I}\sum_{j\in J}\sum_{i'\in I}\sum_{j'\in J} P_{ij}\alpha_{iji'j'}\chi_{ij}\chi_{i'j'} & & \\
\text{Subject to} & \sum_{i\in I}\chi_{ij} = 1, & \forall j\in J, & \text{(M 3.8)} \\
 & \sum_{j\in J}\chi_{ij} = 1, & \forall i\in I, & \\
 & \chi_{ij}\in\{0,1\}, & \forall i\in I, j\in J. &
\end{array}
$$

The quadratic assignment problem (Koopmans and Beckmann 1957)

[7]Note that if i and j are neutral about all pairs of $p \in I$ and $q \in J$, their IP is equivalent to their initial preference.

$$\begin{array}{lll}
\text{Minimize} & \sum_{i \in I} \sum_{j \in J} \sum_{i' \in I} \sum_{j' \in J} f_{ii'} d_{jj'} \chi_{ij} \chi_{i'j'} & \\
\text{Subject to} & \sum_{i \in I} \chi_{ij} = 1, & \forall j \in J, \\
& \sum_{j \in J} \chi_{ij} = 1, & \forall i \in I, \\
& \chi_{ij} \in \{0, 1\}, & \forall i \in I, j \in J.
\end{array}$$

3.3.4.2 Many-to-One Matching with *IP*

Another type of *IP* belongs to many-to-one matching processes between sets I and J, where the elements of set I influence each other's preferences if they all are matched to the same element of set J (see, e.g., Fig. 2.4). In such cases, the mutual *IP* between $i \in I$ and $j \in J$ is calculated using the following formula:

$$\widehat{P}_{ij} = P_{ij}\left(1 + \sum_{i' \in I} \alpha_{ii'} \chi_{i'j}\right), \quad \forall i \in I, j \in J, \tag{3.12}$$

where

- $\alpha_{ii'} > 0$, if i is altruistic about i'.
- $\alpha_{ii'} < 0$, if i is envious about i'.
- $\alpha_{ii'} = 0$, if i is neutral about i'.

Given the above definition, the many-to-one matching process with *IP* is formulated as follows:

$$\begin{array}{lll}
& M : 1/IP/+, OS & \\
\text{Maximize} & \sum_{i \in I} \sum_{j \in J} P_{ij} \chi_{ij} + \sum_{i \in I} \sum_{j \in J} \sum_{i' \in I} P_{ij} \alpha_{ii'} \chi_{ij} \chi_{i'j} & \\
\text{Subject to} & \sum_{i \in I} \chi_{ij} \le M_j, & \forall j \in J, \\
& \sum_{j \in J} \chi_{ij} = 1, & \forall i \in I, \\
& \chi_{ij} \in \{0, 1\}, & \forall i \in I, j \in J.
\end{array} \tag{M 3.9}$$

Model (M 3.9) is an extended version of Model (M 3.2) (i.e., many-to-one matching without *IP*), where the initial preferences are replaced by the *IP* formulated in Eq. (3.12).

Interdependencies between preferences may take other forms—it may not necessarily be a linear function of mutual influences of individuals on each other's preferences. The common characteristic of all variants of matching with *IP*, however, is in the nature of such interdependencies—that the mutual preference of two individuals may be changed, if another matching takes place. The layered matching (*LM*) presented in the next section is another example of interdependent matches, where the quality of match in the current layer is a function of how individuals were matched in the previous layer(s).

3.3.5 Layered Matching

Layered matching (*LM*) refers to sequential matching processes, where the outcomes of preceding matching processes are matched in subsequent stages. Matching subassembly parts is an example of this condition (see Fig. 2.5), where the process of matching four subassembly parts (i.e., sets I, J, K, and L) results in two parts (i.e., sets Y and Z) to be matched to each other in the next stage. In a layered matching process, the mutual preference of individuals at the current layer is a function of the matching outcomes of the previous layers. In the process of matching subassembly parts, for example, the mutual preference of $y \in Y$ and $z \in Z$ is formulated as follows:

$$P_{yz} = \sum_{i \in I} \sum_{j \in J} \sum_{k \in K} \sum_{l \in L} \pi_{ijkl}^{yz} \chi_{ij} \chi_{kl}, \quad \forall y \in Y, z \in Z. \tag{3.13}$$

In Eq. (3.13), parameters π_{ijkl}^{yz} add up (for all i, j, k, and l, where $\chi_{ij} = \chi_{kl} = 1$) to construct P_{yz}. Given the above definition, the two-layer one-to-one matching between sets Y and Z can be formulated as follows:

$$\begin{array}{lll}
 & 1:1/LM/+,OS & \\
\text{Maximize} & \sum_{i \in I} \sum_{j \in J} \sum_{k \in K} \sum_{l \in L} \sum_{y \in Y} \sum_{z \in Z} \pi_{ijkl}^{yz} \chi_{ij} \chi_{kl} \chi_{yz} & \\
\text{Subject to} & \sum_{i \in I} \chi_{ij} = 1, & \forall j \in J, \\
 & \sum_{j \in J} \chi_{ij} = 1, & \forall i \in I, \\
 & \sum_{k \in K} \chi_{kl} = 1, & \forall l \in L, \\
 & \sum_{l \in L} \chi_{kl} = 1, & \forall k \in K, \\
 & \sum_{y \in Y} \chi_{yz} = 1, & \forall z \in Z, \\
 & \sum_{z \in Z} \chi_{yz} = 1, & \forall y \in Y, \\
 & \chi_{ij}, \chi_{kl}, \chi_{yz} \in \{0,1\}, & \forall i \in I, j \in J, k \in K, l \in L, y \in Y, z \in Z.
\end{array} \tag{M 3.10}$$

Model (M 3.10) is a *cubic assignment problem*. The objective function maximizes the overall satisfaction degree of the mutual preferences between the elements of sets Y and Z, with respect to the quadratic preference function presented in Eq. (3.13). The first six sets of constraints ensure that the pairwise relationships $I : J$, $K : L$, and $Y : Z$ are all one-to-one (e.g., between the subassembly parts; Fig. 2.5), and the last set of constraints ensures the feasibility of the decision variables. Note that the given formulation is merely for the two-layer one-to-one matching problem. Adding more layers or changing the types of pairwise relationships between individuals may drastically change the formulation of the problem.

3.4 D3. Criteria

In all the matching models described so far, the matching objective is formulated as the overall satisfaction of individuals based on their mutual preferences. This section presents other formulations of matching objectives along with their application examples.[8] All functions are presented for two-sided matching between sets I and J, considering normalized preference scores as matching criterion.[9]

(1) *Overall satisfaction.* The overall satisfaction function, as presented throughout this chapter, considers the product-sum of the individuals' preferences and matching variables (D3 := $+, OS$):

$$\text{Maximize} \sum_{i \in I} \sum_{j \in J} P_{ij} \chi_{ij}. \tag{3.14}$$

(2) *Bottleneck.* The bottleneck function (Ravindran and Ramaswami 1977) maximizes the minimum satisfaction degree of individuals (D3 := $+, BN$):

$$\text{Maximize} \min_{i \in I, j \in J} \{P_{ij} \chi_{ij}\}. \tag{3.15}$$

(3) *Minimum deviation.* Similar to the bottleneck function, the minimum deviation function (Martello et al. 1984) minimizes the gap between the maximum and minimum satisfaction degrees, or the average and minimum satisfaction degrees (D3 := $+, MD$):

$$\text{Minimize} \sum_{i \in I} \sum_{j \in J} \left(P_{ij} - \min_{k \in I, l \in J} \{P_{kl} \chi_{kl}\} \right) \chi_{ij}. \tag{3.16}$$

---

[8] See Chap. 2, Sect. 2.1.3, for detailed description of each criteria function.

[9] As discussed earlier, such preferences may be representatives of various criteria such as cost, time, distance, or geometric fitness. The matching processes with higher dimensions can be formulated in a similar manner.

(4) *Weighted sum*. In case multiple criteria are involved in the matching process,[10] e.g., n different types of preferences P^1_{ij}, P^2_{ij}, …, P^n_{ij}, the weighted sum function assigns certain weights $(w_1 + w_2 + \ldots + w_n = 1)$ to each and formulates the objective function as follows (D3 := + + ⋯ +, *WS*):

$$\text{Minimize} \sum_{i \in I} \sum_{j \in J} \left(w_1 P^1_{ij} + w_2 P^2_{ij} + \cdots + w_n P^n_{ij} \right) \chi_{ij}. \tag{3.17}$$

(5) *Goal programming*. Various types of goal programming exist in literature. In this section, however, only the *preemptive goal programming* technique is briefly explained, and the rest are left as further reading (see, e.g., Schniederjans 1995). Consider the matching process discussed in (4) with n different types of preferences P^1_{ij}, P^2_{ij}, …, P^n_{ij}. For each type of preference, consider a target value of T and an underachievement level of U, and suppose that all preference types are numbered according to their priority levels (i.e., P^1_{ij}: highest priority; P^n_{ij}: lowest priority). The preemptive goal programming technique is then conducted by solving n models in sequence:

$$\begin{aligned}
1:\quad & \text{Minimize} && U_1 \\
& \text{Subject to} && \text{Goal } 1: \sum_{i \in I} \sum_{j \in J} P^1_{ij} \chi_{ij} + U_1 = T_1 \\
& && \text{Goal } 2: \sum_{i \in I} \sum_{j \in J} P^2_{ij} \chi_{ij} + U_2 = T_2 \\
& && \vdots \\
& && \text{Goal } n: \sum_{i \in I} \sum_{j \in J} P^n_{ij} \chi_{ij} + U_n = T_n \\
& && \text{and other system constraints.}
\end{aligned}$$

$$\begin{aligned}
2:\quad & \text{Minimize} && U_2 \\
& \text{Subject to} && \text{Goal } 1: \sum_{i \in I} \sum_{j \in J} P^1_{ij} \chi_{ij} + U_1 = T_1 \\
& && \text{Goal } 2: \sum_{i \in I} \sum_{j \in J} P^2_{ij} \chi_{ij} + U_2 = T_2 \\
& && \vdots \\
& && \text{Goal } n: \sum_{i \in I} \sum_{j \in J} P^n_{ij} \chi_{ij} + U_n = T_n \\
& && \text{other system constraints, and} \\
& && U_1 = \text{Level achieved in Step 1.}
\end{aligned}$$

$\vdots$

[10] Note that all criteria must be normalized or have the same unit to ensure consistency in the model.

$$
\begin{aligned}
n:\quad & \text{Minimize} && U_n \\
& \text{Subject to} && \text{Goal 1}: \sum_{i \in I} \sum_{j \in J} P_{ij}^{1} \chi_{ij} + U_1 = T_1 \\
& && \text{Goal 2}: \sum_{i \in I} \sum_{j \in J} P_{ij}^{2} \chi_{ij} + U_2 = T_2 \\
& && \vdots \\
& && \text{Goal } n: \sum_{i \in I} \sum_{j \in J} P_{ij}^{n} \chi_{ij} + U_n = T_n \\
& && \text{other system constraints, and} \\
& && U_1 = \text{Level achieved in Step 1} \\
& && U_2 = \text{Level achieved in Step 2} \\
& && \vdots \\
& && U_{n-1} = \text{Level achieved in Step } n-1.
\end{aligned}
$$

3.5 D+. Static Versus Dynamic Matching

All the matching models discussed so far are assumed to be static, i.e., they incorporate no changes in the model, neither in the inputs nor in the characteristics of the sets, conditions, or criteria. In some cases, however, the state of the system is not fixed over time. Hence, the additional dimension of the PRISM taxonomy of best matching (i.e., D+) tends to incorporate such time-dependent changes in the state of the system. More specifically, dynamic matching models consider real-time changes in the input parameters (*DI*: dynamic inputs) and the configurations of the matching sets (*ES*: emergent sets) through association/dissociation of individuals to/from the system (see, e.g., Fig. 2.8). In mathematical terms, the time dependency can be incorporated in the model by adding a time factor t to the model. For example, the dynamic one-to-one matching between sets I and J can be formulated as follows:

$$
\begin{aligned}
& && 1:1//+,OS/DI,ES \\
& \text{Maximize} && \sum_{i \in I(t)} \sum_{j \in J(t)} P_{ij}(t) \chi_{ij} \\
& \text{Subject to} && \sum_{i \in I(t)} \chi_{ij} = 1, && \forall j \in J(t), \\
& && \sum_{j \in J(t)} \chi_{ij} = 1, && \forall i \in I(t), \\
& && \chi_{ij} \in \{0,1\}, && \forall i \in I(t), j \in J(t).
\end{aligned}
\tag{M 3.11}
$$

In Model (M 3.11), $P_{ij}(t)$ represents the variable mutual preference of i and j (as a function of time), and $I(t)$ and $J(t)$ represent emergent sets I and J, respectively. Although simple in terms of mathematical representation, dynamic matching is sometimes difficult to solve due to its variable and uncertain nature. This issue is addressed in detail in Chap. 6 of the book, Dynamic and Distributed Matching.

3.6 Summary

This chapter presented the mathematical models of matching processes—from the basics to more advanced processes—according to the PRISM taxonomy of best matching. Formal representation of such models can provide insights for the development of efficient solution approaches, analogical comparison, interpretation of the results, and valid predictions as well as process improvements. Although the matching models presented in this chapter may not be all-inclusive, they provide an effective basis for the formulation of different classes of best matching processes. The following chapters of this book analyze the details and behaviors of different models and provide efficient algorithms and tools for solving them.

Problems

3.1. Apply the principal mathematical modeling checklist to the library of best matching models presented in this chapter.

3.2. "Best matching models are essentially integer or mixed-integer programs."

(a) Explain this statement.
(b) Illustrate it, including formulation, with two examples of best matching from the previous chapters, once for a two set matching problem, and once for a multiple set matching problem.

3.3. Can you think of ways to expand the library of best matching models presented in this chapter?

3.4. For each of the following matching problems, explain the meaning of each variable and each part of the M 3.1 model:

(a) Selective assembly of chair seats to chair backs.
(b) Assignment of emergency responders to simultaneous emergency calls.

3.5. For each of the following matching problems, explain the meaning of each variable and each part of the (M 3.2) model:

(a) Assignment of job-orders to one machine.
(b) Selection of trainees for design teams.
(c) Selection of chair-legs for a chair-seat.

3.6. Describe two multi-dimensional matching problems, one with planar and one with axial formulation, and for each, explain the meaning of each variable and each part of the respective models, (M 3.3) and (M 3.4).

3.7. For the second dimension of the PRISM taxonomy, illustrate the meaning of the following conditions with examples:

(a) *RC* (Resource Constraints).
(b) *PR* (Precedence Relations).
(c) *RS* (Resource Sharing).
(d) *IP* (Interdependent Preferences).
(e) *LM* (Layered Matching).

3.8. What additional conditions can be added to the list of conditions in Problem 3.7? Explain your answer with examples.
3.9. Can any of the conditions in Problem 3.7 co-exist in the same matching problem? Explain.
3.10. Explain and illustrate why Model (M 3.2) is a special case of Model (M 3.5).
3.11. Illustrate two matching problems, each with precedence relations, and explain how Model (M 3.6) can be used to formulate them.
3.12. Apply Model (M 3.7) to describe the matching problem with resource (capacity) sharing of airlines that participate in an alliance agreement.
3.13. Using a table, compare Models (M 3.8) and (M 3.9) with Model (M 3.7), and illustrate your comparison with specific matching problems in (a) assembly; (b) food service.
3.14. Explain briefly and illustrate the meaning and influence of IP (Interdependent Preferences) on models of best matching.
3.15. Describe several decisions that you have made, that have been influenced by IP. Can they be mapped to any of the best matching models described in this chapter?
3.16. Compare and contrast between layered matching, and matching with interdependent preferences. Use a table to identify the commonalities and the differences.
3.17. Using a table, compare and contrast Model (M 3.10) with other 1:1 matching models presented earlier in this chapter. Then illustrate your answer with matching examples.
3.18. What is the role of the cubic assignment in Model (M 3.10), and why is it needed?
3.19. Each matching problem seeks to satisfy one criterion or several criteria.

(a) Illustrate physical (e.g., geometric), financial, ecological, social, and geographic criteria.
(b) Explain how they can be included in each of the Models (M 3.1), (M 3.6), and (M 3.10).

3.20. What is the main difference between Model (M 3.11) and the previous 10 Models, (M 3.1)–(M 3.10), presented in this chapter? When is this difference meaningful, and when is it not meaningful? Illustrate with two different examples.
3.21. Consider the following general matching problem: Components are moving on a conveyor and arriving at a robot (or several robots); the robot(s) then need to pick these components up and place them in bins (or mobile carts). (Imagine luggage handling in an airport; fruits handling in a packing facility;

instruments handling in a hospital; etc.) Apply each of the 11 models defined in this chapter to specific variations of this general example, and explain clearly your assumptions for each of them.

3.22. Repeat Problem 3.21 for a general matching problem in your area of responsibility.

References

Cobelli, C., & Carson, E. (2001). An introduction to modelling methodology. In E. Carson & C. Cobelli (Eds.) *Modeling methodology for physiology and medicine* (pp. 1–13).

Daskalai, S., Birbas, T., & Housos, E. (2004). An integer programming formulation for a case study in university timetabling. *European Journal of Operational Research, 153*, 117–135.

Dell'Amico, M., & Martello, S. (1997). The k-cardinality assignment problem. *Discrete Applied Mathematics, 76*, 103–121.

Gavish, B., & Pirkul, H. (1991). Algorithms for the multi-resource assignment problem. *Management Science, 37*, 695–713.

Gilbert, K. C., Hofstra, R. B., & Bisgrove, R. (1987). An algorithm for a class of three-dimensional assignment problems arising in scheduling operations. *Institute of Industrial Engineers Transactions, 19*, 29–33.

Gilbert, K. C., & Hofstra, R. B. (1988). Multidimensional assignment problems. *Decision Sciences, 19*, 306–321.

Glover, F., Hultz, J., & Klingman, D. (1979). Improved computer based planning techniques—Part II. *Interfaces, 9*, 12–20.

Koopmans, T. C., & Beckmann, M. (1957). Assignment problems and the location of economic activities. *Econometrica, 25*, 53–76.

Martello, S., Pulleyblank, W. R., Toth, P., & de Werra, D. (1984). Balanced optimization problems. *Operations Research Letters, 3*, 275–278.

Moghaddam, M., & Nof, S. Y. (2014). Combined demand and capacity sharing with best matching decisions in enterprise collaboration. *International Journal of Production Economics, 148*, 93–109.

Moghaddam, M., & Nof, S. Y. (2015). Best matching with interdependent preferences—implications for capacitated cluster formation and evolution. *Decision Support Systems*. Forthcoming.

Öncan, T. (2007). A survey of the generalized assignment problem and its applications. *INFOR, 45*, 123–141.

Ravindran, A., & Ramaswami, V. (1977). On the bottleneck assignment problem. *Journal of Optimization Theory and Applications, 21*, 451–458.

Schniederjans, M. J. (1995). Goal programming: Methodology and applications. USA: Springer, eBook ISBN 978-1-4615-2229-4.

Simaria, A. S., & Vilarinho, P. M. (2004). A genetic algorithm based approach to the mixed-model assembly line balancing problem of type II. *Computers and Industrial Engineering, 47*, 391–407.

Yang, S., & Allenby, G. M. (2003). Modeling interdependent consumer preferences. *Journal of Marketing Research, 40*, 282–294.

Chapter 4
Distributed Decision-Making and Best Matching

Abstract This chapter presents fundamental definitions of distributed systems of operations and decisions, the trade-off between collaborative and competitive behaviors in such systems, and different system architectures in terms of distribution of decision-making authorities. Several examples and illustrations are presented to show the implications of those concepts in the context of best matching. The purpose is to demonstrate the concept with the notion of distributed decision-making and best matching as well as the nature of interactions in such environments.

4.1 Single Versus Multiple Decision-Makers

The previous chapters of this book introduced the notion of best matching in distributed and collaborative/competitive systems, and presented fundamental frameworks and models for understanding and representation of various best matching processes. The goal was to take the first step towards the design of best matching processes through clear definition of their characteristics, requirements, and objectives. It was shown—using several examples and illustrations—that best matching processes are not domain-specific, and play substantial roles in the design and operation of any natural or artificial system that involves *interaction*. The natural next step after identification and specification of best matching problems is solving them, and that is the goal of the second part of this book. Before presenting the mechanics of solution methodologies in terms of best matching algorithms (Chap. 5) and protocols (Chap. 6), this chapter discusses the semantics and the nature of algorithms and protocols, in general, and the context of their applications with respect to the structure of decisional abilities and nature of interactions between individuals in a distributed system.

The term *algorithm* originates from "algoritmi", a Latin form of "Al-Khwarizmi", which is the name of a Persian mathematician, astronomer, and

---

The original version of this chapter was revised: For detailed information please see erratum. The erratum to this chapter is available at 10.1007/978-3-319-46070-3_9

M. Moghaddam and S.Y. Nof, *Best Matching Theory & Applications*, Automation, Collaboration, & E-Services 3, DOI 10.1007/978-3-319-46070-3_4

geographer (780–850 AD) who presented the first systematic solution of linear and quadratic equations in *The Compendious Book on Calculation by Completion and Balancing*. In Mathematics and Computer Science, the notion of algorithm has been defined in a wide variety of ways. In the words of Boolos and Jeffrey, the authors of *Computability and Logic* in 1974,

> No human being can write fast enough, or long enough, or small enough (smaller and smaller without limit ...you'd be trying to write on molecules, on atoms, on electrons) to list all members of an enumerably infinite set by writing out their names, one after another, in some notation. But humans can do something equally useful, in the case of certain enumerably infinite sets: They can give explicit instructions for determining the nth member of the set, for arbitrary finite n. Such instructions are to be given quite explicitly, in a form in which they could be followed by a computing machine, or by a human who is capable of carrying out only very elementary operations on symbols.

In simple words, an algorithm can be defined as *a systematic sequence of operations, expressed in a formal language, for solving a problem*. This definition is sufficient for representing all instances of the term algorithm throughout this book. The purpose of this chapter, however, is not to merely define the notion of algorithms, but to distinguish between them and protocols in terms of definition, context, and applications.

The term *protocol*, according to Merriam-Webster dictionaries, is defined as "a system of rules that explain the correct conduct and procedures to be followed in formal situations". In our terminology, a protocol is *a system of rules* for individuals in a network to effectively interact, be productive, and achieve their goals. In telecommunications, for example, a protocol (*a.k.a.*, communication protocol) refers to *a system of rules* that govern communications and information exchange between two or more individuals. In a multi-agent system, as another example, a protocol (e.g., contract net protocol or task administration protocol[1]) implies *a system of rules* for organizing the relationships and interactions in a distributed network of heterogeneous agents. The general purpose of all protocols, regardless of their application domains, is to allocate and administer a certain set of tasks throughout a distributed network of resources.

A question after defining algorithms and protocols is *how do they differ from each other*? The answer to this question depends on both the distribution of decisional abilities and the nature of interactions.

- *Case 1. Centrally controlled system.* In a centrally controlled system with a single decision-maker, the *system of rules* for administering that system can be defined in terms of a *systematic sequence of operations* for the sole decision-maker. In this case, no distinction can be made between an algorithm and a protocol.
- *Case 2. Distributed system involving multiple decision-makers.* In a distributed system involving multiple decision-makers with various types of relationships

[1] *Contract Net Protocol (CNP)* was developed by Smith in the 1970s. *Task Administration Protocols (TAP)* were developed by Nof in the 1990s.

(hierarchical, heterarchical, or a combination[2]), each decision-maker most likely follows a different *sequence of operations*, and thus, the *system of rules* for administering that system (i.e., the protocol) comprises two or more algorithms. Some roles of these algorithms will include coordination among the multiple decision-makers.

Other cases with multiple decision-makers; e.g., centrally controlled system with multiple decision-makers; centrally controlled system with multiple centrally controlled systems, each by a single decision-maker, will also require protocols. Figure 4.1 presents a schematic contextual comparison between algorithms and protocols, which will serve as a basis for the rest of discussions in this and the next two chapters.[3] As shown, as long as there are multiple decision-makers (controllers), whether they are in competition or collaboration, there must be certain level of coordination, which already requires one or more protocols. For a centralized system, when there is just a single decision-maker (controller), there is no need for a protocol, and algorithms are sufficient. Further discussion follows.

4.2 Distribution of Decisional Abilities

Decision-making processes, in general, are composed of four main stages (Fischhoff and Johnson 1990): (1) Identification of all possible courses of action/inaction; (2) Evaluation of possible outcomes; (3) Assessment of the risk and likelihood of each possible outcome; and (4) Integration of insights, rationalization, and decision-making. This sequence of decision-making processes may be handled by one or multiple decision-makers (Fig. 4.2). In a team of mobile robots deployed for material handling in a warehouse or crop monitoring in an agricultural field, for example, the decisions regarding communication, information sharing, movements, or task allocation may be made by one central decision-maker, by a set of cluster leaders, or by individual autonomous robots. Each approach with specific mechanism of interaction and delegation of authority, however, has its own advantages and limitations.

Decision-making is a logical process of making a "choice" out of a pool of alternatives; e.g., deciding *who does what*, *how*, and *when*, in a distributed network of individuals. The logical process supporting a decision is indeed based on an algorithm—simple, complex, or somewhere in between—and is administered by a decision-maker. Two principal factors in a decision-making process are the responsibilities and objectives of the decision-maker, based on which he/she/it processes and makes a decision. Traditional system administration approaches view

---

[2]These concepts will be further elaborated in this chapter.

[3]**Disclaimer**: The terms algorithm and protocol have been used interchangeably in literature, and the terminologies defined and applied by the authors in this book may be defined differently in other sources.

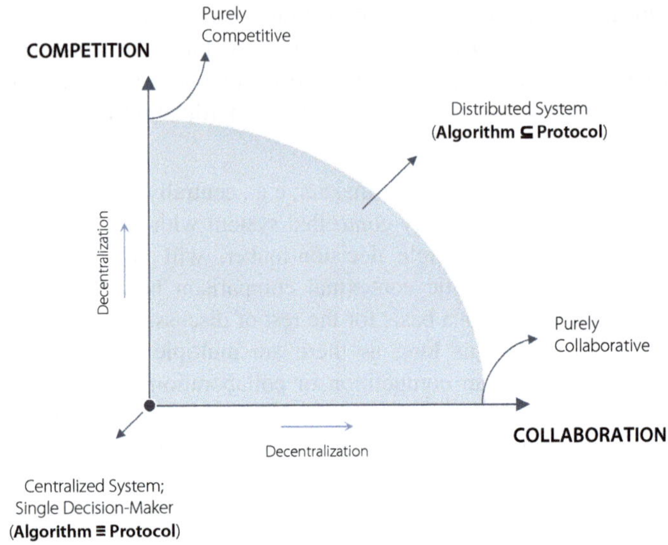

Fig. 4.1 Nature of interactions in distributed systems—contextual comparison between algorithms and protocols

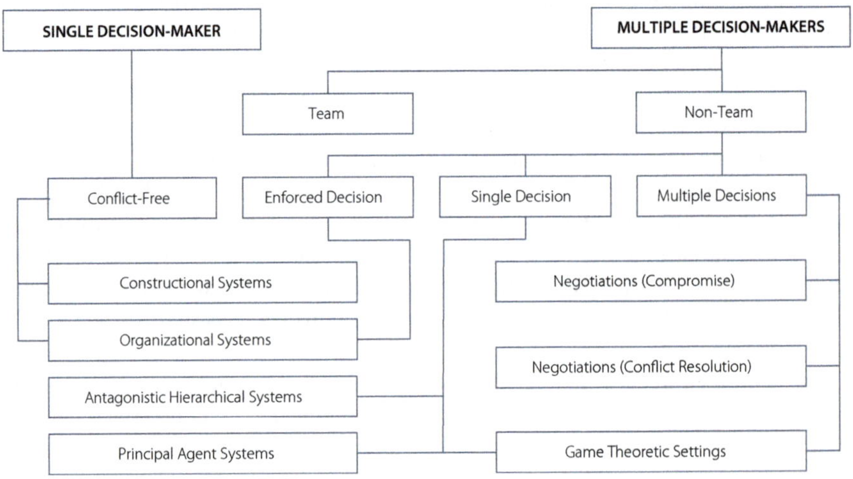

Fig. 4.2 Taxonomy of distributed decision-making. *Constructional systems* team and information symmetry; *Organizational systems* team and information asymmetry. *Antagonistic hierarchical systems* non-team and information symmetry. *Principal agent systems* non-team and information asymmetry (after Schneeweiss 2003)

a system[4] as a *monolithic* and intractably indivisible entity, where all its elements operate under supervision of a central decision-maker (controller). The evolution of systems and the increasing inclination toward *distribution of decisional capabilities*, however, have brought up new challenges to systems engineers and thinkers. With more than one decision-maker, systems have to be prepared for handling the escalating collaborative and competitive interactions among distributed decision-makers, and the emerging complexities associated with them.

Over the past two decades, the world has witnessed drastic increases in the demand for higher efficiency, fault-tolerant, resilient, adaptable, scalable, and dynamic control architectures (Putnik et al. 2013; Nof et al. 2015). In line with the developments in information, computer, and communication technologies, classic monolithic systems have gradually transformed into distributed and complex networks of individuals that need to interact to jointly create value and achieve individual and common goals. Agent-based technologies have enabled the distribution of intelligence, autonomy, and decision-making authority among networks of homogenous or heterogeneous agents (Jennings et al. 1998; Leitão 2009; Barbati et al. 2012); e.g., sensors, controllers, machines, robots, humans, organizations. Overall, each approach has its own advantages and shortcomings. Classic monolithic decision-making, despite inflexibility and inefficiency, results in relatively more optimal decisions and fewer conflicts. Distributed decision-making, on the other hand, diminishes the relative optimality of decisions, and intensifies conflicts, requires negotiations and compromises, in spite of its flexibility, operational efficiency, and delegation of authority.

In a distributed system, the relationship between two individuals is *hierarchical*, *heterarchical*, or a combination of both. Hierarchy implies master-slave relationships, while heterarchy refers to situations where individuals are *unranked* and may have different forms of relationships. Unlike classic hierarchical systems with rigid multi-level master-slave relationships, individuals in a heterarchical system have mutual accessibility to each other and independent modes of operation. The increased frequency of interactions, communications, and conflicts, together with the lack of holistic view of an entire system by individuals and sub-optimality of decisions, however, prevent purely heterarchical systems from absolute acceptability. Accordingly, distributed systems are typically designed with both hierarchical and heterarchical relationships, to alleviate the limitations of pure hierarchy and pure heterarchy while preserving their inherent benefits. That is, while a *range* of decisions is made by a central *mediator*, the rest can be handled autonomously by distributed individuals (Fig. 4.3). The following real-life examples further illuminate the notions of centralized and distributed decision-making, and their implications for the design of best matching processes.

[4]In this context, a system may refer to any manufacturing, agricultural, security, defense, transportation, supply, air traffic control, or healthcare system.

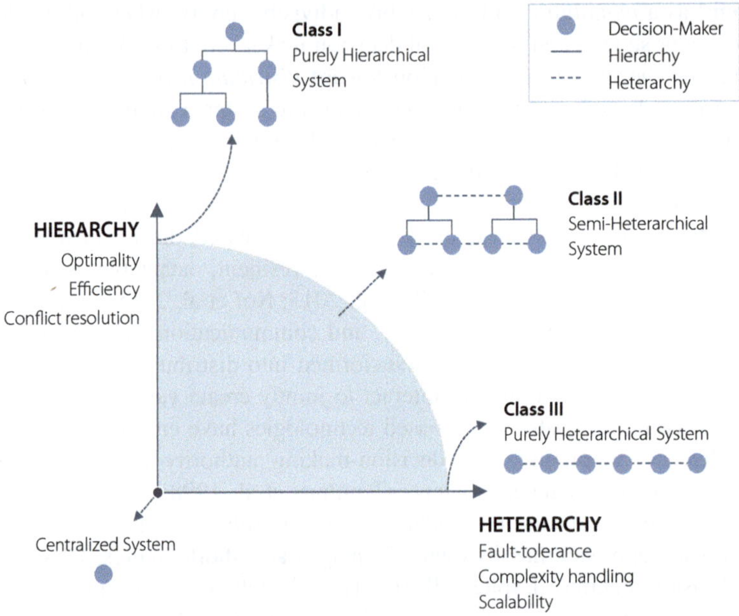

Fig. 4.3 Classification of decision-making abilities. Transition from centralized into a purely hierarchical (*vertical*), purely heterarchical (*horizontal*), and semi-heterarchical (*shaded area*) system (after Trentesaux 2009)

4.2.1 *Example 1: Intelligent Warehouse Management Systems*

The massive growth in the diversity of consumer products provided by online retailers such as Amazon, e-Bay, and Google, along with the increasing demand for faster/more accurate delivery have urged those large e-Commerce corporations to rethink their warehousing operations. Amazon, Inc., for example, is transforming its warehouse operations through automation of fulfillment processes with autonomous robots and AS/RS (Automated Storage/Retrieval System) to achieve higher efficiency and throughput.[5] Such modern warehouses and fulfillment centers deploy a variety of *autonomous* robots for operations such as pick-place and material handling (Fig. 4.4). The emerging warehousing technologies are envisioned to enhance intelligent warehouse management systems with smooth and rapid delivery of products in highly automated environments. In spite of all the developments on mechanical abilities, precision, and intelligence of individual robots, however, realization of those objectives may not be guaranteed without effective design and *orchestration* of the distributed processes.

[5]See www.amazonrobotics.com.

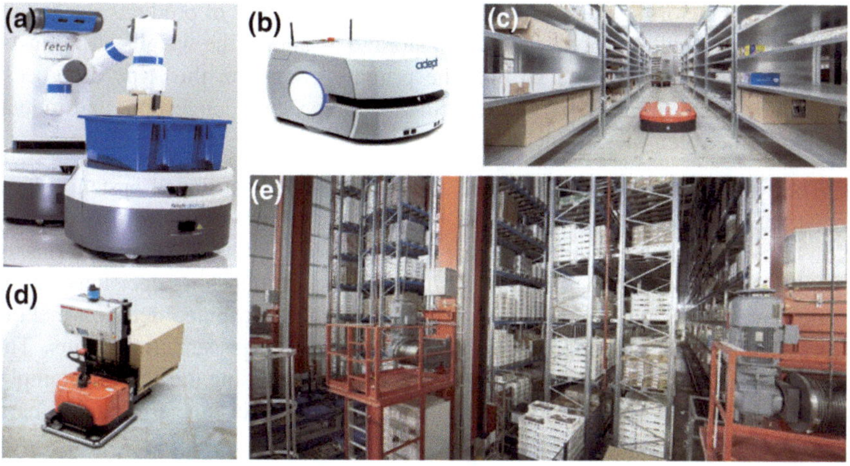

Fig. 4.4 Examples of robotic solutions for intelligent warehousing. **a** Fetch and Freight robots collaborating to fulfill orders (courtesy of Fetch Robotics). **b** Lynx autonomous indoor vehicle for material handling in dynamically changing environments (courtesy of Adept Technology, Inc.). **c** CarryPick Mobile Storage System for carrying intralogistics requirements in fulfillment centers with fluctuating demand (courtesy of Swisslog). **d** The Hybrid Automated Guided Vehicle (AGV) system for cost-efficient pallet transportation (courtesy of Swisslog). **e** AS/RS for constant capacity and careful product handling and storage (courtesy of Vanderlande Industries, Inc.)

Two key factors that significantly influence the performance of warehouse management systems in practice are (1) the design of best matching processes, and (2) the distribution of decisional abilities. Consider, as an example, a warehouse with two types of robot. Type A are material handling robots responsible for carrying products between storage racks and delivery centers (e.g., Fig. 4.4a–d). Type B robots, on the other hand, are AS/RS robots responsible for picking up products from the floor/handling robots and placing them on specific locations on the storage racks, and vice versa (e.g., Fig. 4.4e). The warehouse involves several robots of each type. The primary best matching decisions associate with each type of robots include:

- *Robots Type A*. At each point in time, there are several items to be carried and several robots to carry those items. This is indeed a dynamic best matching process between the set of material handling robots and the set of products, which is influenced by various factors such as real-time location of the robots, their mechanical specifications (e.g., capacity, speed), size and weight of products, and potential priorities (e.g., urgent orders, delicate items). Different criteria may be involved in the design of those best matching processes; e.g., average waiting time, congestion, service rate.
- *Robots Type B*. This type of robots deals with a dynamic matching process between the set of arriving orders/products and the set of storage locations on the rack (see, e.g., Fig. 2.2a). The matching process may be influenced by a

variety of factors such as urgency of orders, the sequence of products stacked on a material handling robot, the size of storage locations, the weight of products, the incoming orders and so on. The primary criteria in this type of process include empty movements of the AS/RS robot, system congestion, and mutual affinity between the products.

The warehouse explained above is an example of a distributed system with a dynamic set of decisions that have to be taken, in some way, on a regular basis. Now the question is what is the best configuration for this system in terms of distribution of decision-making abilities/roles? The best matching decisions can be made centrally, in a hierarchical manner, or through a distributed decision-making process. Each of these approaches has its own advantages and limitations. The answer to this question will be further clarified after discussing the second example, which is about the application of swarm robotics for precision farming and agricultural monitoring.

4.2.2 Example 2: Precision Agriculture

Agriculture is perhaps one of the most important industries providing the most basic human needs—food and fuel. The importance of agriculture industry has drastically escalated over the past few decades, as reported by the Food and Agriculture Organization[6] of the United Nations—globally, over 800 million people suffer from chronic hunger, and around 2 billion suffer from micronutrient deficiencies, while over a third of the global agricultural products is wasted, mostly in early production stages. That highlights the significant impact of the emerging robotics, cyber-physical, and automation solutions for precision crop farming and field monitoring (Fig. 4.5). These technologies focus on collecting farm and field information in order to provide farmers with a solid base for adjusting farming strategies of irrigation, fertilization, disease and weed control, and so on. As differences in farm fields are recognized, farmers can apply optimized, variable control actions, instead of making decisions based on hypothetical average or recent conditions. Due to the inherent cost of precision farming (e.g., facility, installation, manual and automatic control), there is a need for advanced robotics and automation systems that include (1) synchronized movement of robots inside, over, and across fields for collecting information, (2) high-resolution treatment; e.g., weeding and fertilization, in unknown terrains, and (3) interaction with human operators and supervisors in remote stations.

Similar to the warehouse management example, realization of the above goals depends highly on the distribution of decision-making abilities along with the administration of best matching processes. Consider, for example, an agricultural robotic system with two types of robots. Robots type X are drones responsible for

[6]See www.fao.org.

Fig. 4.5 Examples of robotics and automation solutions for precision agriculture. **a** Hornet drone for producing high-resolution images/maps of the field and choosing proper fertilizers (courtesy of Agribotix). **b** 3D Robotics drone for capturing aerial images of the field (courtesy of 3D Robotics). **c** Lancaster drone for capturing humidity, temperature, pressure, and incident light data for crop researchers and farmers (courtesy of PrecisionHawk). **d** 1000 Pruning Robot for autonomous pruning of crops (courtesy of Wall-Ye). **e** BoniRob lightweight swarm robot for weeding, applying fertilizer, and inspection (courtesy of Amazone-Werke Gmbh)

imaging and environmental sensing for features such as humidity and temperature (e.g., Fig. 4.5a–d). Type Y robots are lightweight unmanned ground vehicles designed for tasks such as weeding, pruning, and crop monitoring and inspection. In a large field with several type X and type Y robots, a dilemma—from systems engineering and control perspective—is to effectively design and administer the best matching processes associated with the robots' individual operations along with their interactions with other robots and the environment. In such systems, common best matching decisions include:

- *Robots Type X*. The primary best matching decision to be made by this type of robot is task allocation; i.e., matching a dynamic set of drones and a dynamic set of tasks. The set of drones is dynamic in terms of their availability, due to their fluctuating energy levels and assigned tasks in real-time. The set of tasks is also dynamic, in terms of their quantity and definition—a set of operations can be decomposed into different numbers of tasks with different intensities/durations (e.g., covering specified regions, sensing certain types of crops, measuring various features). The matching process must be conducted with respect to specific conditions, such as real-time location of robots, their physical specifications and capabilities (e.g., speed, energy consumption, battery level), and certain properties of the agricultural field (e.g., types of crops, dimensions, wind speed/direction). The criteria for this type of matching processes can be cost, energy consumption, coverage rate, processing time, and/or conflict rate. The latter is related to the design of interaction mechanisms between drones as well as with the ground robots, in case multiple robots are needed for processing a set of tasks. In such cases, other best matching processes may be involved, including matching robots to each other/certain teams, considering their compatibilities in terms of physical characteristics and real-time status.

- *Robots Type Y.* The best matching processes for ground robots are similar to those of drones. At each point in time, different types of tasks (e.g., weeding, inspection, applying fertilizers) arrive, each relating to specific regions of the field, which must be assigned to the best ground robots available, considering their real-time status (e.g., load, battery level, location, current tasks, maintenance schedule). These matching processes are typically designed in a way that criteria such as average processing time, coverage rate, energy consumption, or total cost of the system are optimized. Another matching process associated with the operation of ground robots is related to lateral collaboration with other robots through data exchange, task sharing, or collaborative processing of the same set of tasks (e.g., BoniRob robots; Fig. 4.5e). In this context, finding the best peer(s) or team with which to collaborate is indeed a dynamic matching process performed in line with the local objective of the robot (e.g., energy saving) and/or the global objective of the system (e.g., full coverage in the minimum time), and with respect to certain conditions (e.g., operational compatibility of robots).

Similar to the warehouse system of Example 1, the aforementioned agricultural robots indeed form a distributed system with various types of best matching processes and decisions to be handled in real-time. As the size of such systems increases, in terms of number of distributed individuals involved in the process, the distinction between central and distributed decision-making mechanisms becomes more noticeable. The choice of the best configuration for decision-making in distributed systems, however, is questionable, as none of them absolutely dominates the others and each has its own advantages and limitations.

4.2.3 Alternative Configurations—Advantages and Limitations

According to the classification of decisional abilities presented in Fig. 4.3, the best matching decisions associate with robots A, B, X, and Y can be made in one of the following manners:

(1) *Centralized.* In a centralized system, distributed decisions such as pickup-delivery and routing of robots A, storage allocation and movements of robots B, allocation of imaging/sensing tasks to robots X, or assignment of regions to robots Y for weeding/inspection—all are made by one decision-maker (e.g., a program running on a central server). The decisions are

then directed to the (passive) distributed individuals for action. This classic approach, in spite of global optimality of decisions, is not sufficiently scalable, fault tolerant, and capable of handling large-scale processes. High frequencies of communication between the central server and the distributed, often remote robots reduce the attractiveness of this configuration in terms of cost-effectiveness and reliability. Centrally controlled systems are unreliable, because failure of the central server implies failure of the entire system.

(2) *Hierarchical*. An alternative to the classic centralized system is a hierarchical system, where the respective best matching decisions are processed by multiple decision-makers organized in different layers (e.g., master-slave relationships between the central server and local controllers on each robot). This approach is beneficial in terms of robustness, predictability, and conflict resolution. The *bureaucratic* decision-making process, however, diminishes the responsiveness, fault tolerance, and scalability of hierarchical systems. In addition, similar to the centralized system, hierarchical relationships lead to lower reliability, because failure of a high-level individual most likely results in failure of low-level individuals.

(3) *Heterarchical*. Heterarchy implies no hierarchy—in a heterarchical system, each pair of individuals can dynamically take over the roles of *master* and *slave*. More heterarchy is equivalent to further delegation of decision-making roles to the distributed individuals. Accordingly, in a purely-heterarchical warehouse or precision agriculture system, for example, each individual robot is responsible for making its own local decisions and sharing its knowledge with other peers if required, in order to enhance their decision-making processes. This configuration resolves the limitations of centralized and hierarchical systems in terms of lack of scalability, responsiveness, fault-tolerance, and complexity handling capabilities. A critical downside of heterarchical configurations, however, is that the individual decision-makers have merely local knowledge of the system's status and requirements, and make decisions based on their own perspective/objectives. This, in turn, impairs global optimality of the decisions. In addition, full delegation of decisions to distributed individuals (e.g., robots type A, B, X, or Y) is not yet technologically feasible.

(4) *Semi-heterarchical*. Unlike the three configurations described above, a semi-heterarchical system does not have any *crisp* description. It sweeps through a wide range of alternative configurations—from purely hierarchical to purely heterarchical (see Fig. 4.3). Accordingly, the best distribution of decision-making roles is indeed an additional decision to be made based on the requirements and objectives of the system. *Team formation* is a common solution for distributed robotic systems under semi-heterarchical configuration. In the warehouse example, robots type A can form multiple teams, where the best matching decisions associated with each team are made by the team leader, which can be one of the robots in the team or an external controller (i.e., hierarchical relationship between the leader and team members). Each team can also collaborate with other teams by sharing information, knowledge, tasks, and/or members (i.e., heterarchical relationship between teams/leaders).

Similar mechanisms can be applied for arrangement of other types of robots and their decision-making processes. Semi-heterarchical configuration is intended to resolve all the limitations of the other three configurations while preserving their advantages as much as possible. The challenge, however, is to find the best combination of hierarchy and heterarchy that materializes this goal in a distributed system such as a warehouse or an agricultural field.

4.3 Nature of Interactions

As discussed in Chap. 1, interaction or *reciprocal influence* of individuals (e.g., genes, people, organizations, countries) on each other range between two different behaviors: collaboration, and competition (see Fig. 4.1). The main distinction between these two behaviors is found in the compatibility of individual and common objectives—more commonality leads to higher level of collaboration, and vice versa. This property is universal. In natural organisms, for example, the commonality among individuals is represented in terms of their *relatedness*—individuals treat their relatives more altruistically, because they share a high proportion of their genes (see, e.g., West et al. 2002). In a supply network, as another example, the commonality may be defined in terms of the *portfolio of market* associated with each enterprise, and the *degree of overlap* between those portfolios (Khanna et al. 1998). Axelrod and Hamilton (1981) exemplify this concept by an example from international trade, where two nations have imposed trade barriers on each other's exports:

> Because of the mutual advantages of free trade, both countries would be better off if these barriers were eliminated. But if either country were to eliminate its barriers unilaterally, it would find itself facing terms of trade that hurt its own economy. In fact, whatever one country does, the other country is better off retaining its own trade barriers. Therefore, the problem is that each country has an incentive to retain trade barriers, leading to a worse outcome than would have been possible had both countries cooperated with each other.

Nature favors collaboration, at least *within* teams, groups, organizations, or societies. In spite of competitions between neighbors in a society (e.g., genes for transmission; honeybees for foraging roles; siblings for family resources), it is scientifically proven that a key factor for the evolution of large and complex societies in nature is *repression of competition*. That is, instead of competitions, individuals increase their benefit by increasing the efficiency of their society through collaboration (*a.k.a., fair meiosis*). Such collaborative behaviors—founded on *durability* of interactions rather than trust—drive the evolution of societies and cannot be overcome by selfish behaviors of destructive individuals (Axelrod and Hamilton 1981). In the words of Frank (2003),

> Fairness aligns individual interests with group interests; aligned interests enhance group efficiency; efficiency promotes group success in competition against other groups; thus, group-against-group competition favors fairness within groups.

This rule is also generalizable to artificial systems. In a supply network, for example, the competition within strategic groups is much less intensive than the competition between strategic groups; because higher levels of commonality and mutual dependence between the members of each group (Caves and Porter 1977). In spite of the benefits of collaboration, competition is an inevitable behavioral pattern in almost every society, from microorganisms to multinational corporations. Altruistic and selfish behaviors always appear side by side and give each other relative meaning. The notion of "coopetition" is coined in game theory to represent situations where competitors prefer to both compete and collaborate with each other (see Bengtsson and Kock 2000). This behavior corresponds with the shaded area of Fig. 4.1.

To clarify this concept, consider the famous "Prisoner's Dilemma" game. The game is about two members of a criminal gang who are arrested, and have to respond to prosecutors with no means of communicating with the other prisoner. The prosecutors do not have enough evidence to convict both prisoners, and therefore offer them a bargain: Each prisoner is given an opportunity to either betray (i.e., compete with) the other prisoner by testifying that the other committed the crime, or collaborate with the other prisoner by remaining silent (loyal to each other). If they both betray each other, each will serve two years in prison. If only one betrays, he/she will be set free but the other one will serve three years in prison. If none of them betrays, each of them will serve one year in prison. At first glance, competition (by betrayal) seems to be a better approach, as it always yields a better result. If the other prisoner betrays, betrayal leads to a two-year sentence, while collaboration (by him/her staying loyal) results in a three-year sentence (to them). If, on the other hand, the other prisoner collaborates (stays loyal to the other prisoner), betrayal leads to freedom, while collaboration by both (staying loyal to each other) results in a one-year sentence to both. The dilemma is that if both collaborate (with each other), the outcome will be much better than the case where they both betray (see Fig. 4.6).

The Prisoner's Dilemma demonstrates the interdependence influence of individual's behaviors on the local and global outcomes of the system, as well as their influence on the interaction mechanisms in the future. Direct reciprocity is a simple example of such influences, where individuals collaborate in the first move, and then proceed based on whatever the other player did in the previous move (see, e.g., Fig. 1.4). As explained earlier and shown in Fig. 4.1, however, the nature of interactions is neither purely collaborative nor purely competitive—it is indeed "coopetitive" in most cases (Bengtsson and Kock 2000). That is, while individuals attempt to collaborate to some extent (e.g., conducting R&D activities in collaboration with competitors), they also care about their competitiveness in areas where there is a low degrees of commonality (e.g., launching a new product, unique to each competitor). This attitude implies that the type of interactions is not binary (i.e. , either collaborative or competitive), and instead covers a wide range of combinations of competition and collaboration.

The concept of coopetition can be generalized to the Prisoner's Dilemma game by considering a series of questions (instead of one) asked from each prisoner,

Fig. 4.6 The Prisoner's Dilemma (courtesy of RAND Corporation) [*: with each other]

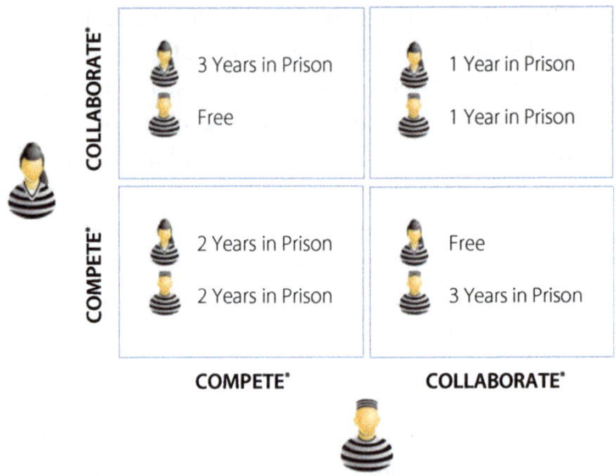

where they can choose between collaboration and betrayal several times by picking an answer for each question. In this case, the outcome is not necessarily 100 % collaboration or betrayal, and rather is a combination of both with certain levels of intensity. The optimal level of intensity and portfolio of such collaborative-competitive behaviors, however, is still an open multidisciplinary question. The most remarkable consensus in most related areas of study is that collaboration is the dominating behavior within societies, while competition is a decisive behavior in interaction with other societies (Alexander and Borgia 1978). With this argument, another challenging question arises about the definition and optimal formation of a society. In the warehouse management and precision agriculture systems discussed in the previous sections, for example, the robots can be members of dynamic teams (i.e., societies), which may compete internally (e.g., for saving energy) or collaborate with other teams (e.g., to enhance the overall productivity). These and several similar challenging questions indicate the complexity of interactions in distributed systems, and the need for effective algorithms and protocols for their design and administrations in practice.

4.4 Summary

This chapter summarized the fundamental concepts and theories of distributed decision-making. It was shown that unlike central decision-making processes in traditional operation research and management approaches, distributed decision-making is influenced by two major factors: (1) distribution of decisional abilities, and (2) nature of interactions. As discussed, the extreme configurations and mechanisms (e.g., purely hierarchical/heterarchical or competitive/collaborative) are not practical/optimal, and the preferred design approaches recommend combinations of them; i.e., semi-heterarchical configurations, coopetitive behaviors. These findings introduce a new cohort of challenges in terms of design complexity and uncertainty. In terms of system configuration, the main challenges center around finding the best portfolio of hierarchical and heterarchical relationships that yield the desirable optimality of decisions, efficiency, fault tolerance, scalability, as well as capability of handling complexity and resolving conflicts. Those design challenges are followed by several questions on the nature of interactions in a society: Do collaborative/competitive behaviors provide competitive advantages in a society? Do such behaviors influence the competitiveness of individuals in the future? What are the advantages/limitations of coopetitive behaviors, and their impacts on interactions with other individuals? When and why should an individual join/leave a society, or change the role? Why would an individual opt to maintain a collaborative/competitive interaction? What are the criteria for assessing the outcomes of a collaborative/competitive behavior? What are the asymmetries (e.g., power, resources, social network) and paranoid concerns (e.g., fear of losing one's identity, mutual stereotyping)? How do collaborative relationships (e.g., trust, mutual learning) evolve? What are the strategies for dealing with uncertainties, openness, emergence, and complexity of coopetitive interaction? The next two chapters provide fundamental solutions and guidelines for addressing such challenges in distributed systems, with reference to the design and control of best matching processes.

Problems

4.1. Consider joggers and walkers sharing a three-lane exercise track. To enable safe use of this track and prevent collisions:

  (a) What algorithm should each participant apply?
  (b) Which task administration protocol is necessary?

4.2. Explain your answers for the following cases regarding the example in Problem 4.1:

  (a) All participants apply an identical algorithm.
  (b) Each of them applies a different algorithm.
  (c) They change their identical algorithm to another identical algorithm based on the other participants present on the track.
  (d) They change their individual, different algorithms based on the other participants present on the track.

4.3. What matching problems can you identify in Problem 4.1? Which of them involve decision making that is: (i) centralized? (ii) decentralized? (iii) competitive? (iv) collaborative? (v) coopetitive?

4.4. Classify the different matching processes of Problem 4.1 according to the taxonomy presented in Fig. 4.2.

4.5. Repeat Problems 4.1–4.4 for the following matching cases:

(a) Drones sharing air-space with passenger airplanes near an airport.
(b) A crew of three robotic farm harvesters having to harvest a large field.
(c) A team of 60 rescuers tasked to rescue people lost in an avalanche.

4.6. Explain the differences between the three classes of systems defined in Fig. 4.3, and illustrate your answer in the cases of

(a) Intelligent warehouse management.
(b) Precision agriculture.

4.7. Describe the two factors that influence matching in the case of intelligent warehouse management, and illustrate how they determine the choice of model from the library of matching models defined in Chap. 3.

4.8. How would the two factors that influence matching in the case of intelligent warehouse management apply in the case of precision agriculture?

4.9. How would Robots Type A and Robots Type B fit into the two types of best matching processes associated with the other robot type?

4.10. Apply three different matching models (Chap. 3) to:

(a) Robots Type A.
(b) Robots Type Y.
(c) A combination of Type X and Type Y robots.

4.11. Define in the context of interaction:

(a) Reciprocal influence.
(b) Team formation.
(c) Relatedness.
(d) Degree of overlap
(e) Orchestration.
(f) Affinity.
(g) Interdependent preference.

Explain how each of the concepts above relates to best matching by using particular models from the library defined in Chap. 3.

4.12. Classify the four cases of Fig. 4.6 according to:

(a) Nature of interactions (Fig. 4.1).
(b) Taxonomy of distributed decision making (Fig. 4.2).
(c) System classes (Fig. 4.3).
(d) Matching models (Chap. 3).

References

Axelrod, R., & Hamilton, W. D. (1981). The evolution of cooperation. *Science, 211*, 1390–1396.

Alexander, R.D., & Borgia, G. (1978). Group selection, altruism, and the levels of organization of life. *Annul Reviews of Ecology and Systematics*, *9*, 449–474.

Barbati, M., Bruno, G., & Genovese, A. (2012). Applications of agent-based models for optimization problems: A literature review. *Expert Systems with Applications, 39*, 6020–6028.

Bengtsson, M., & Kock, S. (2000). "Coopetition" in business networks—To cooperate and compete simultaneously. *Industrial Marketing Management, 29*, 411–426.

Boolos, G., & Jeffrey, R. (1999). *[1974] Computability and logic* (4th ed.). London: Cambridge University Press. ISBN 0-521-20402-X.

Caves, R. E., & Porter, M. E. (1977). From entry barriers to mobility barriers: conjectural decisions and contrived deterrence to new competition. *The Quarterly Journal of Economics, 91*, 241–262.

Fischhoff, B., & Johnson, S. (1990). The possibility of distributed decision-making. In *Distributed Decision Making: Report of a Workshop*. Committee on Human Factors, National Research Council. ISBN: 0-309-58249-0.

Frank, S. A. (2003). Perspective: Repression of competition and the evolution of cooperation. *Evolution, 57*, 693–705.

Jennings, N. R., Sycara, K., & Wooldridge, M. (1998). A roadmap of agent research and development. *Autonomous Agents and Multi-Agent Systems, 1*, 275–306.

Khanna, T., Gulati, R., & Nohria, N. (1998). The dynamics of learning alliances: Competition, cooperation, and relative scope. *Strategic Management Journal, 19*, 193–210.

Leitão, P. (2009). Agent-based distributed manufacturing control: A state-of-the-art survey. *Engineering Applications of Artificial Intelligence, 22*, 979–991.

Nof, S. Y., Ceroni, J., Jeong, W., & Moghaddam, M. (2015). Revolutionizing collaboration through e-work, e-business, and e-service. Springer, ISBN 978-3-662-45777-1.

Putnik, G., Sluga, A., ElMaraghy, H., Teti, R., Koren, Y., Tolio, T., et al. (2013). Scalability in manufacturing systems design and operation: State-of-the-art and future developments roadmap. *CIRP Annals - Manufacturing Technology, 62*, 751–774.

Schneeweiss, C. (2003). *Distributed Decision Making*. Springer-Verlag Berlin Heidelberg.

Trentesaux, D. (2009). Distributed control of production systems. *Engineering Applications of Artificial Intelligence, 22*, 971–978.

West, S. A., Pen, I., & Griffin, A. S. (2002). Cooperation and competition between relatives. *Science, 296*, 72–75.

Chapter 5
Static and Centralized Matching

Abstract This chapter presents the popular algorithms for optimization of static and centralized best matching problems, along with detailed numerical examples and illustrations from various manufacturing, supply, and service systems. The purpose is to introduce and exemplify different exact, heuristic, metaheuristic algorithms, and enable the development of new algorithms based on the standard algorithms introduced in this chapter for handling the computational complexity of different matching processes. The purpose is to explain the role of such algorithms in the context of best matching in systems, and their impact on systems' value and performance.

5.1 Motivation for Using Algorithms

Best matching processes can be designed and administered by several basic algorithms. The algorithms discussed in this chapter are essentially *optimization algorithms*, with systematic sequences of operations that are executed *iteratively* to find the optimal match. According to the taxonomy of interaction mechanisms and the related comparison between algorithms and protocols presented in Fig. 4.1, the best matching algorithms in this chapter are centralized decision-making procedures, executed and controlled by a single decision-maker. The decision-making process therefore involves neither hierarchical/heterarchical relationships, nor collaborative/competitive behaviors, as there is only one necessary decision-maker.[1] Various types of algorithms are introduced for solving different instances of best matching problems, following the PRISM taxonomy of best matching (see Chap. 3). Before discussing the mechanics of best matching algorithms in detail, however, two points need to be clarified:

(1) The goal is to present the core algorithms for centralized best matching. They can be extended and developed further by users for solving special instances

[1]The mechanisms for distributed decision-making with hierarchical/heterarchical relationships and collaborative/competitive behaviors are presented next in Chap. 6, Best Matching Protocols.

M. Moghaddam and S.Y. Nof, *Best Matching Theory & Applications*, Automation, Collaboration, & E-Services 3, DOI 10.1007/978-3-319-46070-3_5

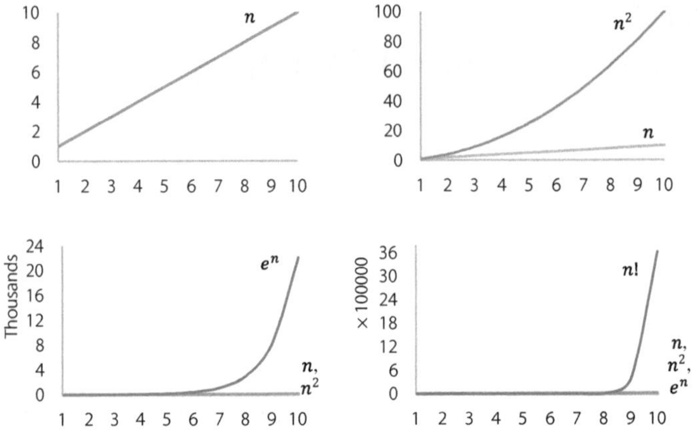

Fig. 5.1 Comparison between constant, quadratic, exponential, and factorial time complexities [x-axis: n]

of best matching problems beyond those covered in this book. Hence, only a small number of exact/approximation algorithms, heuristics, and metaheuristics are discussed here, for the sake of brevity, providing a basis for understanding more advanced and more complicated algorithms.

(2) As the context of this book is application-oriented and multidisciplinary, involving various engineering, management, business, and technology areas in both academia and industry, some theoretical discussions (e.g., computational efficiency of the algorithms) are omitted.[2] Instead, illustrative examples with simple interfaces are presented for better comprehension of the mechanics of operations.

Best matching is known as a computationally expensive problem. Even the simple one-to-one matching, i.e., $1:1//+,OS$ (see Model (M 3.1) in Chap. 3), has quadratic growth $O(n^2)$, as the number of decision variables is equal to n^2, where n denotes the number of individuals in each set—this implies $n!$ feasible solutions (see Fig. 5.1)! Thus, several algorithms have been developed and presented in literature over more than six decades to handle the computational complexity of best matching problems. The rest of this chapter is devoted to the discussion of a selected group of those algorithms, their unique characteristics, advantages, and limitations, and the types of problems they are designed to solve.

[2]Interested readers can read "Matching Theory" by Lovasz and Plummer (1991) and "Assignment Problems" by Burkard et al. (2012).

5.2 Heuristics and Exact Algorithms

An algorithm is called exact if it can find the optimal solution to an optimization problem. Traditional exact optimization procedures are sometimes unable to solve computationally expensive problems, due to their extremely large search spaces (i.e., the set of all feasible points of an optimization problem that satisfy the constraints). In this context, heuristics are proven effective for finding approximate solutions to such "hard" optimization problems by trading optimality, completeness, and accuracy for computational efficiency. Heuristics are usually ad hoc, suitable for particular problem instances, and cannot be easily reused in other application instances—the reason is their high dependency on the knowledge and expertise of the heuristic developer. Higher involvement of intuition and empirical "rules-of-thumb", however, provides heuristic developers with more flexibility to "think out of the box" and create solutions that may be difficult to prove in theory, but are extremely efficient in practice. The next subsections introduce a number of well-known heuristics and exact algorithms for best matching, their history, and logic together with examples and illustrations. It is shown that although heuristics (e.g., Lagrangian relaxation method) are typically faster, exact algorithms (e.g., branch-and-bound method) yield solutions with relatively higher quality.

5.2.1 Hungarian Method

This method is a pioneering and probably the most influential approach towards the development of efficient and practical solutions for best matching. The Hungarian method (Kuhn 1955) is a combinatorial optimization algorithm that solves the original two-sided one-to-one matching problem in polynomial time. The algorithm is based on the duality theorem in linear programming as well as combinatorial tools in graph theory. Kuhn named the algorithm in honor of two Hungarian mathematicians, Dénes König and Jenö Egerváry, whose earlier works provided the basis for the Hungarian method (Kuhn 1955).

Logic

The Hungarian method yields the optimal one-to-one matching between the elements of sets I and J, based on their respective preference matrix P, where P_{ij} denotes the mutual preference of $i \in I$ and $j \in J$ (see Model (M 3.1) in Chap. 3). The method is based on the following theorem: If a number is added to or subtracted from all of the entries of any row or column of a cost matrix, an optimal match for the resulting cost matrix is also an optimal match for the original cost matrix (Kuhn 1955). Since the objective of the original algorithm is to minimize the cost of matching the elements of I and J, we also assume that the entries of the

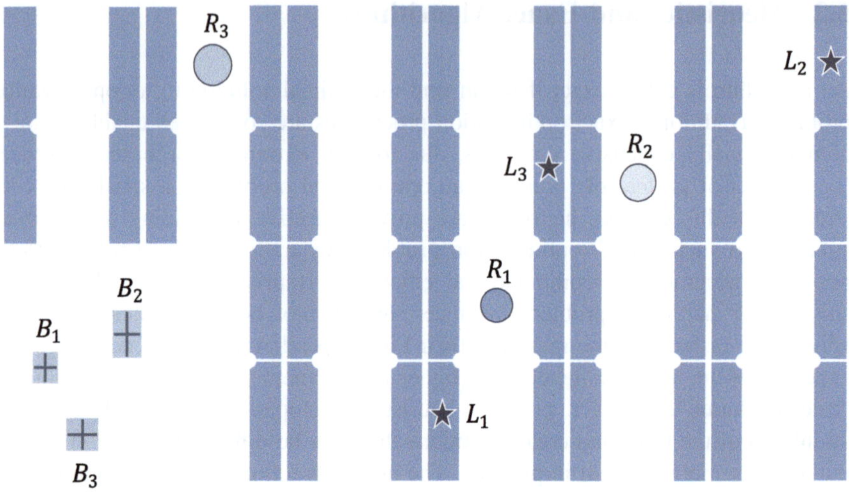

Fig. 5.2 Layout of the warehouse, current locations and specifications of robots and boxes. The robots must move from their current locations to where the boxes are located, pick them up, and carry them to their designated locations in the warehouse [*Stars* locations of the boxes—$L_1 : B_1$; $L_2 : B_2$; $L_3 : B_3$]

preference matrix P are undesirable (e.g., cost) and must be minimized[3] (Note that P is necessarily a square matrix). The Hungarian method is defined as follows:

Step 1 Generate the preference matrix P (with undesirable entries).
Step 2 Subtract the smallest entry of each row from all the entries of that row.
Step 3 Subtract the smallest entry of each column from all the entries of that column.
Step 4 Draw the *minimum* number of vertical and/or horizontal lines that cover all the zero entries of the modified matrix P.
Step 5 If the number of lines is equal to the size of sets I and J, report the zero entries as the optimal matching between I and J. Otherwise, go to Step 6.
Step 6 Identify the smallest entry of the modified matrix P not covered by a line. Subtract that entry from each uncovered row, add it to each uncovered column, and then return to Step 4.

Example

Recall from Chap. 4 the example of intelligent warehouse management systems with two types of robots for handling (Type A) and automated storage/retrieval (Type B) of products in a warehouse (see Fig. 4.4). Figure 5.2 shows a case where three boxes B_1, B_2, and B_3 have arrived in the warehouse and must be carried to specified locations by three material handling robots R_1, R_2, and R_3. (Each box must

[3]The best matching problem to solve is $1:1//-, OS$. A maximization problem can be simply converted to this format by reversing the preference values.

Table 5.1 Movement times of boxes by robots (min)

Robots	Boxes		
	B_1	B_2	B_3
R_1	3.5	5.0	4.0
R_2	3.0	4.5	2.5
R_3	2.0	4.0	3.0

be carried by exactly one robot.) Table 5.1 summarizes the approximate times associated with carrying each box to its predefined location by certain robots, based on the real-time locations and specifications of the robots as well as the weights, dimensions, and destinations of the boxes in the warehouse. The problem is to find the optimal allocation of boxes to robots such that the total movement time of the robots is minimized.

This is a two-sided one-to-one best matching problem (see Model (M 3.1) in Chap. 3) between sets $I = \{R_1, R_2, R_3\}$ and $J = \{B_1, B_2, B_3\}$, with the objective of minimizing the total movement time, i.e., $\sum_{i \in I} \sum_{j \in J} P_{ij} \chi_{ij}$, where

$$\chi_{ij} = \begin{cases} 1, & \text{if robot } i \text{ carries box } j \\ 0, & \text{otherwise} \end{cases}, \quad \forall i \in I, j \in J.$$

The Hungarian method solves this problem as follows:

Step 1 The preference matrix:

$$P = \begin{bmatrix} 3.5 & 5.0 & 4.0 \\ 3.0 & 4.5 & 2.5 \\ 2.0 & 4.0 & 3.0 \end{bmatrix}.$$

Step 2 Subtract 3.5, 2.5, and 2.0 from Rows 1, 2, and 3, respectively:

$$P \leftarrow \begin{bmatrix} \mathbf{3.5} & 5.0 & 4.0 \\ 3.0 & 4.5 & \mathbf{2.5} \\ \mathbf{2.0} & 4.0 & 3.0 \end{bmatrix} - \begin{bmatrix} 3.5 & 3.5 & 3.5 \\ 2.5 & 2.5 & 2.5 \\ 2.0 & 2.0 & 2.0 \end{bmatrix} = \begin{bmatrix} 0 & 1.5 & 0.5 \\ 0.5 & 2.0 & 0 \\ 0 & 2.0 & 1.0 \end{bmatrix}.$$

Step 3 Subtract 0, 1.5, and 0 from Columns 1, 2, and 3, respectively:

$$P \leftarrow \begin{bmatrix} \mathbf{0} & \mathbf{1.5} & 0.5 \\ 0.5 & 2.0 & \mathbf{0} \\ \mathbf{0} & 2.0 & 1.0 \end{bmatrix} - \begin{bmatrix} 0 & 1.5 & 0 \\ 0 & 1.5 & 0 \\ 0 & 1.5 & 0 \end{bmatrix} = \begin{bmatrix} 0 & 0 & 0.5 \\ 0.5 & 0.5 & 0 \\ 0 & 0.5 & 1.0 \end{bmatrix}.$$

Step 4 The minimum number of lines that cover all the zero entries can be drawn as follows:

$$\begin{bmatrix} 0 & 0 & 0.5 \\ 0.5 & 0.5 & 0 \\ 0 & 0.5 & 1.0 \end{bmatrix}.$$

Step 5 Since the number of lines is equal to the number of robots/boxes ($|I| = |J| = 3$), the zero entries represent the optimal matching, which is R_1 and B_2, R_2 and B_3, R_3 and B_1, with the total movement time of 9.5 min

$$\begin{bmatrix} 0 & \textcircled{0} & 0.5 \\ 0.5 & 0.5 & \textcircled{0} \\ \textcircled{0} & 0.5 & 1.0 \end{bmatrix} \equiv \begin{bmatrix} 3.5 & \textcircled{5.0} & 4.0 \\ 3.0 & 4.5 & \textcircled{2.5} \\ \textcircled{2.0} & 4.0 & 3.0 \end{bmatrix}.$$

5.2.2 *Deferred Acceptance Algorithm*

Soon after the invention of the Hungarian method, David Gale and Lloyd S. Shapley, two mathematicians from Brown University and the RAND Corporation (at the time), investigated the same two-sided one-to-one matching problem, but from a somewhat different perspective. In their initial work "College Admissions and the Stability of Marriage", Gale and Shapley (1962) developed an algorithm that does not necessarily yield the optimal matching with respect to the mutual preferences of individuals, but guarantees *stable matching*. According to the Gale-Shapley definition of stability,

> a matching between sets I and J is stable if (1) no individual from either set is paired with an unacceptable match from the other set, and (2) there is no unmatched pair of individuals who both prefer each other to their current matches.

The seminal work of Gale and Shapley was continued by several researchers for centralized market design, with applications such as matching interns to hospitals, students to colleges, and human organs to recipients (Roth 2008). In 2012, the Royal Swedish Academy of Sciences awarded the Nobel Prize in Economic

Sciences to Lloyd S. Shapley and Alvin E. Roth for the theory of stable matching and the practice of market design.

Logic

The deferred acceptance algorithm is suitable for finding the stable one-to-one or one-to-many match[4] between the elements of sets I and J in polynomial time. The algorithm is based on an iterative centralized negotiation mechanism, where the elements of one set offer matching proposals in each iteration, and the elements of the other set respond by either accepting, holding, or rejecting those proposals. The proposal offering and evaluation mechanisms are based on the preferences of the individuals for each other. The preference matrix P is therefore defined slightly differently with entries $P_{ij} = \left(p_{i \to j}, p_{j \to i}\right)$, where $p_{i \to j}$ and $p_{j \to i}$ denote the normalized preferences $i \in I$ for $j \in J$ and $j \in J$ for $i \in I$, respectively. In view of that, the algorithm can be executed in two different manners considering who is proposing (either elements of I or elements of J), which may result in completely different matching results.[5] The deferred acceptance algorithm for one-to-one matching of sets I and J, with the elements of I proposing, is as follows:

Step 1 Each $i \in I$ proposes to its most preferred $j \in J$ (with maximum $p_{i \to j}$).

Step 2 Each $j \in J$ rejects all the unacceptable proposals, and then, if there is more than one acceptable proposal, *holds* the most preferred one (with maximum $p_{j \to i}$) and rejects the others.

Step 3 If all $i \in I$ are matched, the stable match is achieved and the algorithm stops. Otherwise, each $i \in I$ rejected in Step 2 proposes to its next most preferred $j \in J$, and the algorithm returns to Step 2.

The algorithm can be extended, with minor modifications, for solving one-to-many matching problems. No matter who is proposing, the iterative negotiation process must be conducted in a way that (1) the capacity limits are not violated, (2) the proposers are allowed to propose until reaching their limit, and (3) the responders are allowed to reconsider and//or hold proposals for evaluation in the next iterations (see, e.g., Roth 2008). Note that although the negotiation process is handled by distributed individuals, the decision-making process is considered centralized as all individuals follow the exact same rules and procedures defined by one algorithm.

---

[4]The best matching problems to solve are $1:1//+,ST$ and $M:1//+,ST$, where ST denotes stability.

[5]The deferred acceptance algorithm sacrifices optimality for stability; i.e., the resulting match does not necessarily maximize the overall satisfaction of the individuals' preferences. In addition, it can be shown that the elements of the proposing set are generally more satisfied compared to the elements of the responding set, which poses some questions on the fairness of the algorithm. For market design applications, however, the deferred acceptance algorithm has been highly practical and successful.

Example
Recall the example of the collaborative community healthcare network discussed in Chap. 2 (Fig. 2.3). Suppose that each hospital is going to hire exactly one of the doctors, and thus, the doctors are asked to submit applications for those jobs. Recasting the application and hiring process in the context of the deferred acceptance algorithm, this case can be regarded as a matching between sets $I = \{D_1, D_2, D_3, D_4\}$ (doctors) and $J = \{H_1, H_2, H_3, H_4\}$ (hospitals) with the elements of I proposing (The doctors apply and the hospitals respond). This process is based on the preferences of the doctors and the hospitals for each other (Fig. 5.3), formalized between 1 and 4 (1: least preferred; 4: most preferred) as follows:

$$P = \begin{bmatrix} (2,2) & (1,2) & (4,1) & (3,4) \\ (1,4) & (3,1) & (2,3) & (4,3) \\ (1,1) & (3,3) & (4,4) & (2,2) \\ (3,3) & (1,4) & (2,2) & (4,1) \end{bmatrix},$$

where in $P_{ij} = (a, b)$, a and b denote the preference of i for j and the preference of j for i, respectively. The deferred acceptance algorithm with the doctors proposing finds the stable match as follows:

Step 1 Doctors D_1, D_2, D_3, and D_4 submit their first applications to Hospitals H_3, H_4, H_3, and H_4, respectively

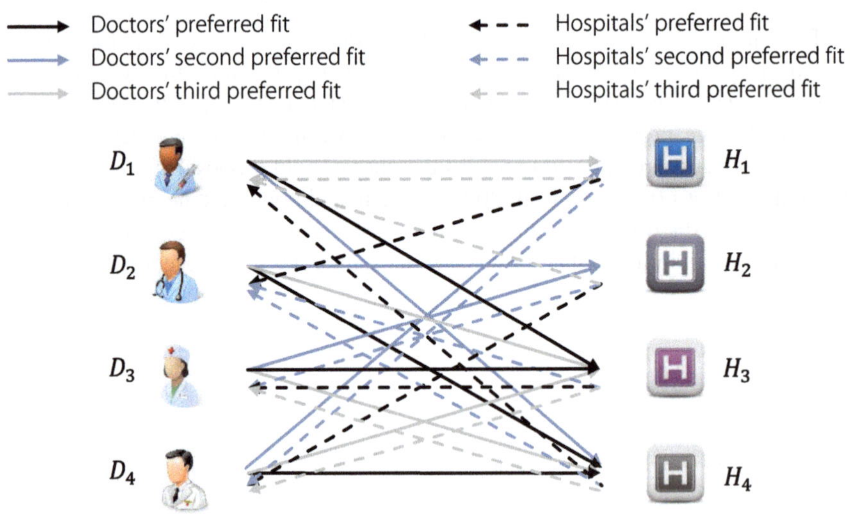

Fig. 5.3 Matching doctors and hospitals: using the deferred acceptance algorithm, the matching outcomes will be different depending on who makes the offer and who responds (e.g., when doctors offer, the respective stable match is $\{D_1, H_4; D_2, H_2; D_3, H_3; D_4, H_1\}$)

$$D_1 \rightarrow H_3$$
$$D_2 \rightarrow H_4$$
$$D_3 \rightarrow H_3$$
$$D_4 \rightarrow H_4$$

Step 2 Hospital H_3 that has two applications holds D_3 and rejects D_1, because $p_{H_3 \rightarrow D_3} = 4 > 1 = p_{H_3 \rightarrow D_1}$. Similarly, Hospital H_4 holds D_2 and rejects D_4

D_{1} \to H_{3} ✗
D_{2} \to H_{4} ✓
D_{3} \to H_{3} ✓
D_{4} \to H_{4} ✗

Step 3 Since Doctors D_1 and D_4 are unmatched, they submit their applications to their second preferred hospitals, i.e., H_4 and H_1, respectively

$$D_1 \rightarrow H_4$$
$$D_4 \rightarrow H_1$$

Step 2 (*Revisited*) Hospital H_1 holds D_4 as its only applicant. Hospital H_4 rejects its current match, i.e., D_2, as the new application is more preferred $(p_{H_4 \rightarrow D_1} = 4 > 3 = p_{H_4 \rightarrow D_2})$

D_{1} \to H_{4} ✓
D_{2} \to H_{4} ✗
D_{3} \to H_{3} ✓
D_{4} \to H_{1} ✓

Step 3 (*Revisited*) D_2 is the only unmatched doctor. However, his second preferred hospital, i.e., H_2 has luckily not hired anybody yet. Hence, he submits the application, and since he is the only applicant of Hospital H_2, his application is accepted, and the algorithm stops with the following stable matching decision:

D_{1} \to H_{4} ✓
D_{2} \to H_{2} ✓
D_{3} \to H_{3} ✓
D_{4} \to H_{1} ✓

5.2.3 *Lagrangian Relaxation Method*

A substantial number of computationally expensive optimization problems can be viewed as “easy” problems complicated by a small number of constraints

(Fisher 1981). The Lagrangian relaxation method (Held and Karp 1970, 1971) is a smart use of this property for solving "hard" constrained optimization problems through relaxation of certain sets of constraints—those that cause the computational complexity. The idea is to provide useful information through approximation by replacing strict inequalities with penalty costs associated with the violation of those inequalities using Lagrange multipliers. The Lagrangian relaxation method is widely studied for solving the generalized matching problem with resource constraints (Öncan 2007). The solutions to the relaxed problem provide suitable (upper/lower) bounds to the original problem, which can then be used for solving the original problem using iterative Lagrangian relaxation method, or other methods such as branch-and-bound (see Sect. 5.3).

Logic

Recall the resource constrained generalized matching problem discussed in Chap. 3 [see Model (M 3.5)]:

$$\text{Maximize} \quad Z = \sum_{i \in I} \sum_{j \in J} P_{ij} \chi_{ij} \tag{5.1}$$

$$\text{Subject to} \quad \sum_{i \in I} r_{ij} \chi_{ij} \le l_j, \quad \forall j \in J, \tag{5.2}$$

$$\sum_{j \in J} \chi_{ij} = 1, \quad \forall i \in I, \tag{5.3}$$

$$\chi_{ij} \in \{0, 1\}, \quad \forall i \in I, j \in J. \tag{5.4}$$

Two Lagrangian relaxations can essentially be applied to the above model (Fisher 1981). The first relaxation (L1) associated with dualizing (i.e., relaxing) Constraints (5.2) is easier to solve, compared to the second relaxation (L2), which dualizes Constraints (5.3). The second relaxation, however, provides better bounds compared to the first relaxation.

L1: Relaxation of Constraints (5.2)

The first relaxation is obtained by dualizing Constraints (5.2), and adding them as a new term to Objective Function (5.1) along with their respective Lagrange multipliers $\lambda_j \in \mathbb{R}_+^{|J|}$, $\forall j \in J$, as follows:

$$Z_{L1} = \sum_{i \in I} \sum_{j \in J} P_{ij} \chi_{ij} - \sum_{j \in J} \lambda_j \left(\sum_{i \in I} r_{ij} \chi_{ij} - l_j \right). \tag{5.5}$$

The modified model can be formulated as

$$\begin{aligned} \text{Maximize} \quad & Z_{L1} = \sum_{i\in I}\sum_{j\in J}\left(P_{ij} - \lambda_j r_{ij}\right)\chi_{ij} + \sum_{j\in J}\lambda_j l_j \\ \text{Subject to} \quad & \sum_{j\in J}\chi_{ij} = 1, \quad \forall i \in I, \\ & \chi_{ij} \in \{0,1\}, \quad \forall i \in I, j \in J. \end{aligned} \tag{L1}$$

The solution to relaxation (L1) can be easily obtained by calculating the optimal values of the decision variables using the following set of equations:

$$\chi_{ij}^{*} = \begin{cases} 1, & \text{if } P_{ij} - \lambda_j r_{ij} = \max_{k\in J}\{P_{ik} - \lambda_k r_{ik}\} \\ 0, & \text{otherwise} \end{cases}, \quad \forall i \in I, j \in J.$$

L2: Relaxation of Constraints (5.3)

The second Lagrangian relaxation can be done in a similar manner, through dualizing Constraints (5.3) and adding them to Objective Function (5.1) with Lagrange multipliers $\mu_i \in \mathbb{R}_{+}^{|I|}$, $\forall i \in I$, as follows[6]:

$$Z_{L2} = \sum_{i\in I}\sum_{j\in J} P_{ij}\chi_{ij} - \sum_{i\in I}\mu_i\left(\sum_{j\in J}\chi_{ij} - 1\right). \tag{5.6}$$

The modified model can be formulated as

$$\begin{aligned} \text{Maximize} \quad & Z_{L2} = \sum_{i\in I}\sum_{j\in J}\left(P_{ij} - \mu_i\right)\chi_{ij} + \sum_{i\in I}\mu_i \\ \text{Subject to} \quad & \sum_{i\in I} r_{ij}\chi_{ij} \le l_j, \quad \forall j \in J, \\ & \chi_{ij} \in \{0,1\}, \quad \forall i \in I, j \in J. \end{aligned} \tag{L2}$$

Model (L2) can be recast as $|J|$ independent "0/1 knapsack problems"[7] (KP), with the *j*th problem formulated as

$$\begin{aligned} \text{Maximize} \quad & Z_{KP_j} = \sum_{i\in I}\hat{P}_{ij}\chi_{ij} \\ \text{Subject to} \quad & \sum_{i\in I} r_{ij}\chi_{ij} \le l_j, \\ & \chi_{ij} \in \{0,1\}, \ \forall i \in I. \end{aligned} \tag{KP$_j$}$$

---

[6]A concern about the use of the Lagrangian relaxation method is to find the best Lagrange multipliers that yield good approximations with solutions close to the optimal solutions of the original problem. Ross and Soland (1975) suggest that the best values for Lagrange multipliers of Model (L2) lie in $P_{i,2} \le \mu_i \le P_{i,1}$, where $P_{i,1}$ and $P_{i,2}$ are the largest and the second largest values of P_{ij} for all $j \in J$. Interested readers are referred to Fisher (1985) and subsequent works.

[7]*Knapsack problem*: Given a set of items each with certain weight and benefit, find a collection of items that yields the maximum benefit without violating a specified weight limit.

where $\hat{P}_{ij} = P_{ij} - \mu_i, \forall i \in I, j \in J$. The optimal solutions to all $|J|$ 0/1 KPs can be obtained through Dynamic programming (Martello et al. 2000), and then integrated to construct the optimal solution to Model (L2) i.e.,

$$Z^*_{\text{L2}} = \sum_{j \in J} Z^*_{\text{KP}_j} + \sum_{i \in I} \mu_i. \tag{5.7}$$

Dynamic Programming for Solving (KP$_j$)

Step 1 *Initialization*. Define auxiliary variable $\theta \in [0, l_j]$ as the *remaining capacity* of the knapsack, and set $Z_{\text{KP}_j}(0, \theta) = 0$ and $Z_{\text{KP}_j}(i, 0) = 0$.

Step 2 *Recursive calculation*. Perform the following recursion from $Z_{\text{KP}_j}(0, 0)$ to $Z_{\text{KP}_j}(|I|, l_j)$ by alternatively increasing the values of i and θ:

$$Z_{\text{KP}_j}(i, \theta) = \begin{cases} Z_{\text{KP}_j}(i-1, \theta), & \text{if } r_{ij} > \theta, \\ \max\{\hat{P}_{ij} + Z_{\text{KP}_j}(i-1, \theta - r_{ij}), Z_{\text{KP}_j}(i-1, \theta)\}, & \text{if } r_{ij} \leq \theta. \end{cases} \tag{5.8}$$

and report $Z^*_{\text{KP}_j} = \max_{i,\theta}\{Z_{\text{KP}_j}(i, \theta)\}$ as the optimal solution to (KP$_j$).

Step 3 *Backtracking the solution*. Set $\theta = l_j$, and repeat for $i = |I|, |I| - 1, \ldots, 1$: If $Z_{\text{KP}_j}(i, \theta) \neq Z_{\text{KP}_j}(i-1, \theta)$, set $\chi_{ij} = 1$ and $\theta \leftarrow \theta - r_{ij}$; otherwise, set $\chi_{ij} = 0$.

Example

The operations manager of a startup construction company has recently signed contracts for three independent but similar construction projects. The company currently has 10 employees with different types/levels of expertise, experience, and salary. The budget dedicated to each project is limited. In addition, due to the competition between the three clients and the associated confidentiality issues, each employee can only work on one project. Based on her knowledge of the employees' compatibility to each project, as well as their estimated cost (Fig. 5.4), the manager is challenged to find the best allocation of her employees to the projects such that the overall performance is maximized while the budget limits are not violated. The problem is indeed a resource constrained two-sided one-to-many best matching between sets $I = 1, \ldots, 10$ and $J = 1, \ldots, 3$. The overall performance of the employees and the budget limits are regarded as the matching criterion (i.e., P_{ij}, $\forall i \in I, j \in J$), and the resource constraints (i.e., l_j, $\forall j \in J$), respectively.

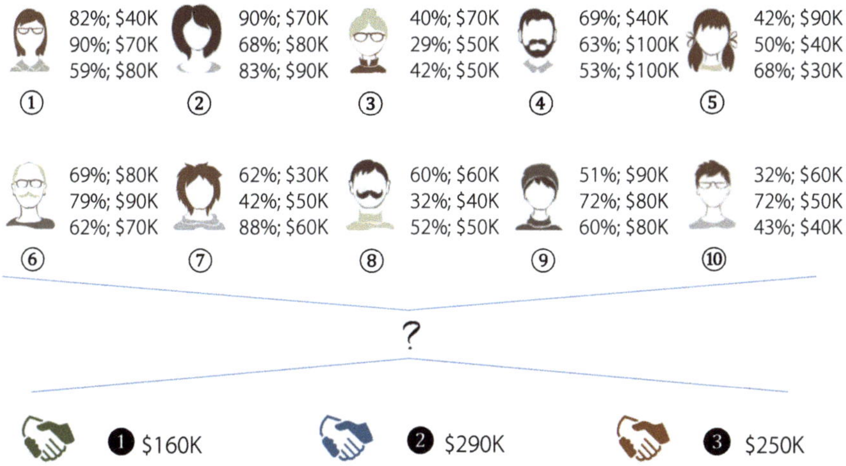

Fig. 5.4 Teaming for construction—manager's projection of the employees' performance and cost, as well as the budgets dedicated to different projects [employee data format: line X ≡ Project X (performance; cost)]

Table 5.2 Solution #1 to the employee and project matching problem (L1)

Project contracts	Employees										Lagrange multipliers
	1	2	3	4	5	6	7	8	9	10	
1		✓		✓				✓			0.40
2	✓					✓			✓	✓	0.30
3			✓		✓		✓				0.50
Z_{L1}	730 ≡ 73.0 % performance on average										
Feasible?	No (violates budget capacity constraints)										

✓: Matched

L1 Results

Tables 5.2 and 5.3 present two different solution sets associated with the first Lagrangian relaxation with two different sets of Lagrange multipliers. The solutions are obtained by plugging each pair of $i \in I$ and $j \in J$ in Eq. (5.6). As shown in this example, the performance of the Lagrangian relaxation method depends significantly on the choice of Lagrange multipliers. Considerable effort is therefore required for selecting proper values that yield solutions Z^*_{L1} that are feasible and reasonably close to Z^* (see Jeet and Kutanoglu 2007). This method, however, provides an efficient mechanism for calculating good bounds for computationally expensive optimization problems with relatively minor effort.

L2 Results

The second Lagrangian relaxation of this problem results in three independent 0/1 KPs. As suggested by Ross and Soland (1975), a good approximation for the

Table 5.3 Solution #2 to the employee and project matching problem (L1)

Project contracts	Employees										Lagrange multipliers
	1	2	3	4	5	6	7	8	9	10	
1	✓	✓		✓							1.10
2						✓			✓	✓	0.80
3			✓		✓		✓	✓			0.90
Z_{L1}	714 ≡ 71.4 % performance on average										
Feasible?	Yes										

✓: Matched

Lagrange multipliers of (L2) is $\mu_i = P_{i,2},\ \forall i \in I$, where $P_{i,2}$ denotes the second largest values of P_{ij} for all $j \in J$. Accordingly,

i	1	2	3	4	5	6	7	8	9	10
μ_i	82	83	40	63	50	69	62	52	60	43

The mathematical formulations of (KP_1), (KP_2), and (KP_3) are presented below.

$$\begin{aligned} \text{Maximize} \quad & Z_{\text{KP}_1} = 7\chi_{2,1} + 6\chi_{4,1} - 8\chi_{5,1} + 8\chi_{8,1} - 9\chi_{9,1} - 11\chi_{10,1} \\ \text{Subject to} \quad & 40\chi_{1,1} + 70\chi_{2,1} + 70\chi_{3,1} + 40\chi_{4,1} + 90\chi_{5,1} + 80\chi_{6,1} \\ & \quad + 30\chi_{7,1} + 60\chi_{8,1} + 90\chi_{9,1} + 60\chi_{10,1} \le 160, \\ & \chi_{i,1} \in \{0,\ 1\},\ \forall i \in I. \end{aligned} \tag{KP$_1$}$$

$$\begin{aligned} \text{Maximize} \quad & Z_{\text{KP}_2} = 8\chi_{1,2} - 15\chi_{2,2} - 11\chi_{3,2} + 10\chi_{6,2} - 20\chi_{7,2} \\ & \quad - 20\chi_{8,2} + 12\chi_{9,2} + 29\chi_{10,2} \\ \text{Subject to} \quad & 70\chi_{1,2} + 80\chi_{2,2} + 50\chi_{3,2} + 100\chi_{4,2} + 40\chi_{5,2} + 90\chi_{6,2} \\ & \quad + 50\chi_{7,2} + 40\chi_{8,2} + 80\chi_{9,2} + 50\chi_{10,2} \le 290, \\ & \chi_{i,2} \in \{0, 1\},\ \forall i \in I. \end{aligned} \tag{KP$_2$}$$

$$\begin{aligned} \text{Maximize} \quad & Z_{\text{KP}_3} = -23\chi_{1,3} + 2\chi_{3,3} - 10\chi_{4,3} + 18\chi_{5,3} - 7\chi_{6,3} + 26\chi_{7,3} \\ \text{Subject to} \quad & 80\chi_{1,3} + 90\chi_{2,3} + 50\chi_{3,3} + 100\chi_{4,3} + 30\chi_{5,3} \\ & \quad + 70\chi_{6,3} + 60\chi_{7,3} + 50\chi_{8,3} + 80\chi_{9,3} + 40\chi_{10,3} \le 290, \\ & \chi_{i,3} \in \{0,\ 1\},\ \forall i \in I. \end{aligned} \tag{KP$_3$}$$

Table 5.4 presents the final matching results obtained by solving the three independent KPs presented above. Note that Employee 4 remained unmatched in all

Table 5.4 Solution to the employee and project matching problem (L2)

Project contracts	Employees									
	1	2	3	4	5	6	7	8	9	10
1		✓						✓		
2	✓					✓			✓	✓
3			✓	✓	✓		✓			
Z_{L2}	714 ≡ 71.4 % performance on average									
Feasible?	Yes									

✓: Matched

Table 5.5 Dynamic programming calculations for (KP_1)[a]

i	θ (×10)																
	0	1	2	3	4	5	6	7	8	9	10	11	12	13	14	15	16
0	0	0	0	0	0	0	0	0	0	0	0	0	0	0	0	0	0
1	0	0	0	0	0	0	0	0	0	0	0	0	0	0	0	0	0
2	0	0	0	0	0	0	0	7	7	7	7	7	7	7	7	7	7
3	0	0	0	0	0	0	0	7	7	7	7	7	7	7	7	7	7
4	0	0	0	0	6	6	6	7	7	7	7	13	13	13	13	13	13
5	0	0	0	0	6	6	6	7	7	7	7	13	13	13	13	13	13
6	0	0	0	0	6	6	6	7	7	7	7	13	13	13	13	13	13
7	0	0	0	0	6	6	6	7	7	7	7	13	13	13	13	13	13
8	0	0	0	0	6	6	8	8	8	8	14	14	14	15	15	15	15
9	0	0	0	0	6	6	8	8	8	8	14	14	14	15	15	15	15
1	0	0	0	0	6	6	8	8	8	8	14	14	14	15	15	15	15

[a]The shaded cells are related to the backtracking stage of the dynamic programming procedure (Step 3) where $\chi_{ij} = 1$

KPs (i.e., $\chi_{4,j} = 0, \forall j \in J$) and thus he was matched to Project Contract 3, as the only one with extra budget capacity.[8] The optimal objective value is calculated using Eq. (5.7). As discussed earlier, relative to the first relaxation (L1), the second relaxation (L2) is more difficult to solve, but is expected to provide better results (compare Z_{L2} in Table 5.4 with Z_{L1} in Tables 5.2 and 5.3). Details of the dynamic programming calculations for (KP_1), (KP_2), and (KP_3) are presented in Tables 5.5, 5.6, and 5.7, respectively.

[8]This is one of the major drawbacks of relaxation L2—solving several KPs in parallel may sometimes result in incomplete/infeasible solutions (e.g., an employee assigned to no project/multiple projects).

Table 5.6 Dynamic programming calculations for (KP_2)[ab]

i	θ (×10)																								
	5	6	7	8	9	10	11	12	13	14	15	16	17	18	19	20	21	22	23	24	25	26	27	28	29
0	0	0	0	0	0	0	0	0	0	0	0	0	0	0	0	0	0	0	0	0	0	0	0	0	0
1	0	0	8	8	8	8	8	8	8	8	8	8	8	8	8	8	8	8	8	8	8	8	8	8	8
2	0	0	8	8	8	8	8	8	8	8	8	8	8	8	8	8	8	8	8	8	8	8	8	8	8
3	0	0	8	8	8	8	8	8	8	8	8	8	8	8	8	8	8	8	8	8	8	8	8	8	8
4	0	0	8	8	8	8	8	8	8	8	8	8	8	8	8	8	8	8	8	8	8	8	8	8	8
5	0	0	8	8	8	8	8	8	8	8	8	8	8	8	8	8	8	8	8	8	8	8	8	8	8
6	0	0	8	8	10	10	10	10	10	10	10	18	18	18	18	18	18	18	18	18	18	18	18	18	18
7	0	0	8	8	10	10	10	10	10	10	10	18	18	18	18	18	18	18	18	18	18	18	18	18	18
8	0	0	8	8	10	10	10	10	10	10	10	18	18	18	18	18	18	18	18	18	18	18	18	18	18
9	0	0	8	12	12	12	12	12	12	12	20	20	22	22	22	22	22	22	22	30	30	30	30	30	30
10	29	29	29	29	29	29	29	37	41	41	41	41	41	41	41	49	49	51	51	51	51	51	51	51	59

[a]The shaded cells are related to the backtracking stage of the dynamic programming procedure (Step 3) where $\chi_{ij} = 1$

[b]The zero columns $(0 \leq \theta \leq 40)$ are eliminated from the table due to space limitation

Table 5.7 Dynamic programming calculations for (KP_3)[a]

i	θ (×10)																									
	0	1	2	3	4	5	6	7	8	9	10	11	12	13	14	15	16	17	18	19	20	21	22	23	24	25
0	0	0	0	0	0	0	0	0	0	0	0	0	0	0	0	0	0	0	0	0	0	0	0	0	0	0
1	0	0	0	0	0	0	0	0	0	0	0	0	0	0	0	0	0	0	0	0	0	0	0	0	0	0
2	0	0	0	0	0	0	0	0	0	0	0	0	0	0	0	0	0	0	0	0	0	0	0	0	0	0
3	0	0	0	0	0	2	2	2	2	2	2	2	2	2	2	2	2	2	2	2	2	2	2	2	2	2
4	0	0	0	0	0	2	2	2	2	2	2	2	2	2	2	2	2	2	2	2	2	2	2	2	2	2
5	0	0	0	18	18	18	18	18	20	20	20	20	20	20	20	20	20	20	20	20	20	20	20	20	20	20
6	0	0	0	18	18	18	18	18	20	20	20	20	20	20	20	20	20	20	20	20	20	20	20	20	20	20
7	0	0	0	18	18	18	26	26	26	44	44	44	44	44	46	46	46	46	46	46	46	46	46	46	46	46
8	0	0	0	18	18	18	26	26	26	44	44	44	44	44	46	46	46	46	46	46	46	46	46	46	46	46
9	0	0	0	18	18	18	26	26	26	44	44	44	44	44	46	46	46	46	46	46	46	46	46	46	46	46
10	0	0	0	18	18	18	26	26	26	44	44	44	44	44	46	46	46	46	46	46	46	46	46	46	46	46

[a]The shaded cells are related to the backtracking stage of the dynamic programming procedure (Step 3) where $\chi_{ij} = 1$

5.2.4 *Branch-and-Bound Method*

Branch-and-bound is a powerful discrete optimization method (Land and Doig 1960). The idea is based on systematic enumeration of candidate solutions through state-space search, enabled by different *branching* and *bounding* strategies. A branch produces two or more candidate solutions with minor but known differences from the current solution. A bound, on the other hand, calculates a lower or upper bound for the objective value that is used for fathoming the candidate solutions. The logical structure of the branching and bounding procedures resembles that of a tree (Dakin 1965). The branch-and-bound method provides an efficient (but computationally exhaustive) mechanism for solving matching problems (despite having exponential worst-case performance). A basic version of this method for solving generalized matching problems with resource constraints (Ross and Soland 1975) is presented in this section.

Logic

Recall the resource constrained generalized matching problem discussed in the previous section [also in Chap. 3, Model (M 3.5)], and name it (OM) (Original Model). The branch-and-bound method for optimization of (OM) is as follows:

Bounding Strategy

Step 1 Relax the capacity constraints of the elements of J to generate the following relaxed model (RM):

$$\begin{aligned} \text{Maximize} \quad & Z_{\text{RM}} = \sum_{i \in I} \sum_{j \in J} P_{ij} \chi_{ij} \\ \text{Subject to} \quad & \sum_{j \in J} \chi_{ij} = 1, \quad \forall i \in I, \\ & \chi_{ij} \in \{0, 1\}, \quad \forall i \in I, j \in J. \end{aligned} \tag{RM}$$

Step 2 Solve (RM) as follows:

$$\chi_{ij} = \begin{cases} 1, & \text{if } P_{ij} = \max\limits_{k \in J} \{P_{ik}\} \\ 0, & \text{otherwise} \end{cases}, \quad \forall i \in I, j \in J, \tag{5.9}$$

and define:

$$I_j = \{i\text{: } \chi_{ij} = 1, i \in I\}, \quad \forall j \in J, \tag{5.10}$$

$$\hat{J} = \left\{ j\text{: } \sum_{i \in I_j} r_{ij} \chi_{ij} > l_j \right\}. \tag{5.11}$$

If $\hat{J} = \emptyset$, stop and report $Z^*_{\text{OM}} = Z^*_{\text{RM}}$. Otherwise, proceed to Step 3.

Step 3 Solve the following KP (Knapsack Problem) for all $j \in \hat{J}$[9]:

$$\begin{aligned} &\text{Minimize} \quad Z_{\mathrm{KP}_j} = \sum_{i \in I_j} G_i \upsilon_{ij} \\ &\text{Subject to} \quad \sum_{i \in I_j} r_{ij} \upsilon_{ij} \geq d_j, \\ &\qquad \upsilon_{ij} \in \{0, 1\}, \forall i \in I_j. \end{aligned} \qquad (\mathrm{KP}_j)$$

where

$$G_i = \min_{k \in J \setminus \{j_i\}} \left\{ P_{ij_i} - P_{ik} \right\}, \quad \forall i \in I, \qquad (5.12)$$

$$d_j = \sum_{i \in I_j} r_{ij} \chi_{ij} - l_j, \quad \forall j \in J, \qquad (5.13)$$

and calculate the upper bound:

$$UB = Z^*_{\mathrm{RM}} - \sum_{j \in J} Z^*_{\mathrm{KP}_j}. \qquad (5.14)$$

Branching Strategy

Step 4 Identify the *separation variable* $\chi_{i^*j^*}$ as the one that maximizes the following formula:

$$f_{i^*j^*} = \max \left\{ \frac{G_i}{r_{ij} / \left(l_j - \sum_{i \in I_j} r_{ij} \chi_{ij} \right)}, i \in I, j \in J, \upsilon^*_{ij} = 0 \right\}. \qquad (5.15)$$

Step 5 Repeat Steps 2 and 3 with fixed values $\chi_{i^*j^*} = 1$ and $\chi_{i^*j^*} = 0$ (separately) for the separation variable, and select the fixed value with larger upper bound. Then return to Step 4 for selecting the next separation variable.[10]

In the branch-and-bound procedure described above, I_j denotes the set of individuals $i \in I$ matched to $j \in J$, and $\hat{J}$ denotes the subset of J with violated capacity. G_i denotes the gap between the preferences of individual $i \in I$ for its most preferred

[9]The Dynamic Programming procedure presented in the previous section can be used for solving these KPs.

[10]Note that the previous separation variables remain fixed and the outcomes of their respective KP(s) are used for identifying the next separation variable.

and second most preferred individuals in set J, and d_j denotes the amount of violation from the capacity limit of $j \in J$. The binary variable υ_{ij} implies that $i \in I$ must be reassigned to its second most preferred $j \in J$, if $\upsilon_{ij}^* = 1$; otherwise, no changes are required.

Example

Recall the example of teaming for construction discussed in the previous section (Fig. 5.4). Three approximate solutions were generated for this problem via two different Lagrangian relaxations of the original model (OM) (see Tables 5.2, 5.3, 5.4). The branch-and-bound method is applied here for computing the optimal solution to this problem through the following steps:

Step 1 The relax model (RM) can be generated using the dataset provided in Fig. 5.4. (The model is omitted for the sake of brevity.)

Step 2 The solution to (RM) is calculated using Eq. (5.9) as follows:

Project contracts	Employees									
	1	2	3	4	5	6	7	8	9	10
1		✓		✓				✓		
2	✓					✓			✓	✓
3			✓		✓		✓			
Z_{L2}	730 ≡ 73.0 % performance on average									
Feasible?	No									

The solution is not feasible; $I_1 = \{2,4,8\}$, $I_2 = \{1,6,9,10\}$, $I_3 = \{3,5,7\}$, and $\hat{J} = \{1\}$.

Step 3 The following KP must therefore be solved for $j = 1$:

$$\begin{aligned} &\text{Minimize} && Z_{\text{KP}_1} = 7\upsilon_{2,1} + 6\upsilon_{4,1} + 8\upsilon_{8,1} \\ &\text{Subject to} && 70\upsilon_{2,1} + 40\upsilon_{4,1} + 60\upsilon_{8,1} \geq 10, \\ &&& \upsilon_{ij} \in \{0,1\},\ \forall i \in I_j. \end{aligned} \qquad (\text{KP}_1)$$

The optimal solution to (KP_1) is $\upsilon_{4.1}^* = 1$ and $\upsilon_{2.1}^* = \upsilon_{8.1}^* = 0$. The upper bound is $UB = 730 - 6 = 724$.

Step 4 The separation variable must be selected between $\chi_{2.1}$ and $\chi_{8.1}$. Using Eq. (5.15), $\chi_{8.1}$ is selected as the separation variable.

Step 5 Repeating Step 2 with the separation variable fixed to $\chi_{8.1}^* = 0$, the following solution is obtained for (RM):

Project contracts	Employees									
	1	2	3	4	5	6	7	8	9	10
1		✓		✓						
2	✓					✓			✓	✓
3			✓		✓		✓	✓		
Z_{L2}	722 ≡ 72.2 % performance on average									
Feasible?	Yes									

Because the obtained solution is feasible, Step 3 is unnecessary. Accordingly, the upper bound for $\chi^*_{8.1} = 0$ is equal to $UB(\chi^*_{8.1} = 0) = 722$. The solution for the separation variable fixed to $\chi^*_{8.1} = 1$ is already calculated above, with the upper bound of $UB(\chi^*_{8.1} = 1) = 724$. Hence, the candidate solutions with $\chi^*_{8.1} = 0$ are fathomed, and $\chi^*_{8.1} = 1$ is considered next.

Step 4 (*Revisited*) The next separation variable is $\chi_{2.1}$.

Step 5 (*Revisited*) Repeating Step 2 with the separation variable fixed to $\chi^*_{2.1} = 0$, the following solution is obtained for (RM):

Project contracts	Employees									
	1	2	3	4	5	6	7	8	9	10
1				✓				✓		
2	✓					✓			✓	✓
3		✓	✓		✓		✓			
Z_{L2}	723 ≡ 72.3 % performance on average									
Feasible?	Yes									

This solution is feasible, and thus Step 3 is omitted. The upper bound is therefore equal to $UB(\chi^*_{2.1} = 0; \chi^*_{8.1} = 1) = 723$. The upper bound with the separation variable fixed to $\chi^*_{2.1} = 1$ is already calculated above, which is equal to $UB(\chi^*_{2.1} = 1; \chi^*_{8.1} = 1) = 724$. Accordingly, the candidate solutions with $\chi^*_{2.1} = 0$ are fathomed, and $\chi^*_{2.1} = \chi^*_{8.1} = 1$ are considered next. This branch, however, will eventually lead to an upper bound smaller than $UB(\chi^*_{2.1} = 0; \chi^*_{8.1} = 1) = 723$, and thus, since the solution presented in the table above is feasible as well, it is reported as the optimal solution (the rest of calculations can be done in a similar manner as described above). Figure 5.5 schematically describes the branch-and-bound procedure for solving the employee and project matching problem.

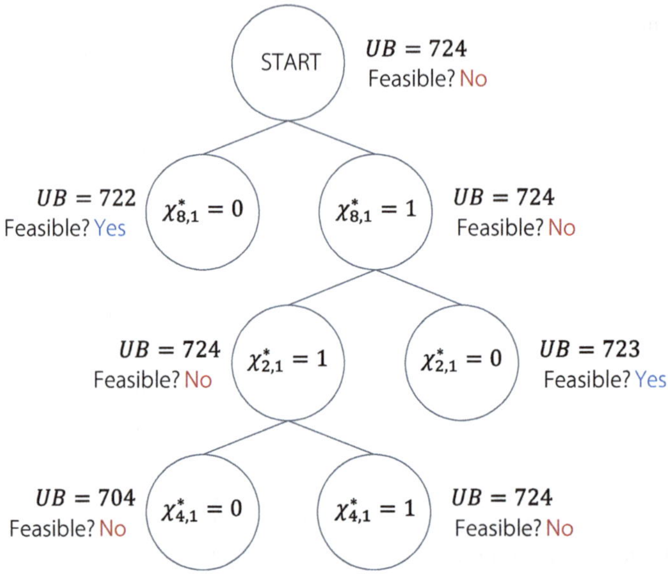

Fig. 5.5 Branch-and-bound tree for the problem of teaming for construction

5.3 Metaheuristics

A metaheuristic is a high-level algorithm for solving optimization problems with unknown structure, incomplete information, and/or limited computational capacity. In the words of Luke (2014), metaheuristics are:

> "algorithms used to find answers to problems when you have very little to help you: you don't know beforehand what the optimal solution looks like, you don't know how to go about finding it in a principled way, you have very little heuristic information to go on, and brute-force search is out of the question because the space is too large. But, if you're given a candidate solution to your problem, you can test it and assess how good it is. That is, you know a good one when you see it."

Metaheuristics are therefore suitable candidates for solving unconventional matching problems, i.e., the ones with unique conditions and/or time-dependent domains on their PRISM taxonomy of best matching (see, e.g., the four examples in Chap. 2, Sect. 2.2). For discrete optimization applications such as best matching, metaheuristics are mainly categorized into three classes:

(1) *Population-based metaheuristics*. This class of metaheuristics finds optimal solutions by iteratively selecting and combining existing alternatives from a set of solutions called "population". The idea behind population-based metaheuristics is inspired by the *evolutionary theory* due to Charles R. Darwin, which states that survival of organisms depends on their differential reproductive success in their interactions with the environment. As implication

for optimization, if each alternative solution is regarded as an individual organism, emulation of the evolutionary mechanisms leads to populations of solutions with better quality. Genetic algorithm (GA)—as the most common instance of population-based metaheuristics—is discussed in Sect. 5.3.1, together with its application for solving best matching problems.

(2) *Constructive metaheuristics*. This class of metaheuristics constructs good solutions through systematic improvement of a partial solution. A common example of constructive metaheuristics is greedy randomized adaptive search procedure (GRASP), which iteratively constructs and improves solutions through greedy heuristic procedures. Swarm intelligence mechanisms are another example; they are built upon the collective behavior of the elements of a society (e.g., ants, termites, wasps, birds, and fish). The implication for optimization is that each member of the emulated swarm contributes to its society by interacting its knowledge with others and enabling a reinforcement learning mechanism that eventually leads to (near-) optimal solutions. GRASP and ACO (Ant Colony Optimization) are two examples of constructive metaheuristics discussed in Sects. 5.3.2 and 5.3.3, and applied for solving best matching problems.

(3) *Local search metaheuristics*. This class of metaheuristics finds optimal solutions by making iterative changes in a *current* solution. Those typically small changes in the current solution are called "neighborhood search", which replaces the current solution with a new and relatively better solution. A popular example of this class of metaheuristics is tabu search, which employs *prohibition* strategies in order to improve the performance of neighborhood search procedures in terms of time and quality. The mechanics of tabu search and its application for solving best matching problems are discussed in Sect. 5.3.4.

5.3.1 Genetic Algorithm *(*GA*)*

GA is a leading metaheuristic pioneered by John H. Holland in the 1970s. The algorithm leads to the survival of the *fittest* members of a population known as "chromosomes", considering their unique properties controlled by their basic units, their "genes". This evolutionary process imitates the process of *natural selection* supported by *crossover* and *mutation* operators, as bio-inspired mechanisms for reproduction and generation of fitter "offsprings". GA can be designed effectively for solving almost all variants of best matching problems, including complicated and advanced instances that are not solvable by the exact and heuristic algorithms discussed thus far (see, e.g., Models (M 3.8) to (M 3.10) in Chap. 3). The challenge, however, is to find the best way to (1) encode a solution set into a chromosome that represents all necessary properties of the problem, and (2) reproduce new populations of chromosomes that lead to the optimal solution in an efficient manner, in terms of both computational time and solution quality. The systematic procedure and rules-of-thumb for designing efficient GAs for best matching problems are described next.[11]

Logic

Although GA must be customized to a given problem instance, the overall procedure for design and implementation of the algorithm for optimization is somewhat generic.[12] A systematic procedure for solving best matching problems using GA is presented below, together with ideas and examples for customization of the algorithm to specific problem instances.[13]

Step 1 *Encoding*. Each potential solution must be encoded as a *chromosome*, which is composed of a set of *genes* with unique *alleles*. Each gene corresponds to a set of decision variables (one or more) and its allele(s) represent their respective value(s). The performance of GA, especially its reproduction operators, depends highly on the selected encoding scheme. For a best matching problem, the most popular chromosomes are in the form of arrays or matrices (Fig. 5.6). Each chromosome must represent a *complete* potential solution, which may or may not be feasible.[14]

Step 2 *Initialization*. The next step after encoding the problem is to generate the initial population of chromosomes. The initial population is the starting point for the evolutionary computation and search for the optimal solution, and thus, its *quality* and *diversity* substantially influence the performance of GA. Below are some common approaches for initialization:

- *Random initialization*. This approach is the simplest (and least efficient). It randomly generates the initial population, following certain conditions. For the team formation example in Fig. 5.6, for instance, each array chromosome must have a length of 10 with integer entries randomly selected from $[1, 3]$. Despite good diversity, this approach does not guarantee acceptable quality or even feasibility of the generated chromosomes.
- *Enhanced initialization*. Certain heuristic or approximation methods can be used to enhance the quality of the initial population, while preserving its diversity. For example, the approximation methods discussed in Sect. 5.2 can be deployed for better initialization; e.g.,

[11]Interested readers are encouraged to read "An Introduction to Genetic Algorithms" by Mitchell (1996) for more in-depth understanding of GA and its applications in other domains.

[12]This property applies to all metaheuristics; and that is the reason they are called "meta"-heuristics.

[13]Note that the examples of two-sided matching are presented for illustration purpose, and all the metaheuristics discussed here can be applied for solving several other (and more complicated) matching problems as well.

[14]The encoding schemes that only allow feasible chromosomes are preferred (if possible to develop) due to their computational efficiency by eliminating infeasible solutions.

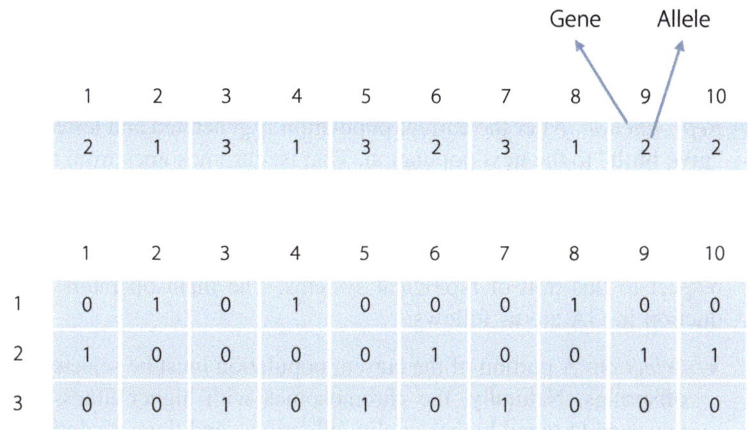

Fig. 5.6 Encoding schemes for the example teaming for construction (Fig. 5.4). Array representation (*top*) and matrix representation (*bottom*) for the same solution set. Array representation is more efficient in terms of computational economy. Other instances of matching can be encoded in a similar fashion—from simple arrays to multidimensional matrices

Lagrangian relaxation with different sets of multipliers, each corresponding to one chromosome of the population. A more efficient initialization approach is GRASP, which is discussed in the next section.[15]

Step 3 *Fitness evaluation*. The natural next step after generation of a population of chromosomes is to evaluate their *fitness*. As quoted by Luke, a good property of metaheuristics like GA is that "*if you're given a candidate solution to your problem, you can test it and assess how good it is.*" Each individual chromosome represents certain values for the decision variables. Hence, calculation of their fitness is a simple task, regardless of the objective function format and criteria (e.g., quadratic, bottleneck, maximum deviation, weighted sum):

- *Objective value*. Plug the values of the decision variables in the objective function and calculate its value. For the solution set in Fig. 5.6, for example, $\chi^*_{1.2} = \chi^*_{2.1} = \chi^*_{3.3} = \chi^*_{4.1} = \chi^*_{5.3} = \chi^*_{6.2} = \chi^*_{7.3} = \chi^*_{8.1} = \chi^*_{9.2} = \chi^*_{10.2} = 1$, and $\chi^*_{ij} = 0$ for other $i \in I$ and $j \in J$. This yields the value of 730 for the objective function.
- *Penalty value*. Plug the values of the decision variables in the constraints. If any of the constraints is violated, add a penalty term to the calculated objective value. In the above solution set, the

[15]Note that initialization approaches are not limited to those discussed here. There are hundreds of initialization approaches in literature, most of which are heuristic and case-based (see E_3 in Chap. 7 for an advanced example).

budget allocated to the first project is \$170, which is more than its budget limit of \$160, and thus, the solution set must be penalized accordingly.[16]

Step 4 *Reproduction*. After the current population is generated and tested, it has to "give birth" to the next population. That is, chromosomes from the current population must combine or mutate, in order to produce *offsprings*. The evolutionary theory implies that natural reproduction must result in a *fitter* population. The reproduction procedure must then be designed with respect to this trait of biological systems. The main operators for reproduction in GA are as follows:

- *Selection*. A portion of the current population must be selected to breed offsprings. Naturally, the chromosomes with higher fitness level are expected to breed higher-quality offsprings, and thus, random selection is not a suitable approach. A popular selection operator is *roulette wheel selection*, which assigns selection probabilities to chromosomes based on their relative fitness values:

$$\Pr_x = \frac{F_x}{\sum_{y=1}^{\Pi} F_y}, \quad \forall x = 1, \ldots, \Pi. \tag{5.16}$$

  where $\Pr_x$ and F_x respectively denote the selection probability and fitness value of chromosome x, and Π is the population size. Selection operators are diverse. Examples of other operators are *stochastic universal sampling*, *tournament selection*, and *Boltzmann selection* (see Mitchell 1996).
- *Crossover*. This is a key reproduction operator in natural selection, which combines the genes of two (or more) parent chromosomes to produce offsprings. The crossover operator must enable probabilistic *exploration* of new solutions (to ensure diversity), and *exploitation* of good traits of parents (to ensure quality). The most basic crossover operator is *point crossover*, which cuts and combines two (or more) parents at randomly selected point(s) (Fig. 5.7a). This operator does not necessarily guarantee the feasibility of offsprings generated from feasible parents. Examples of alternative operators for (possibly) better crossover are *uniform crossover*, *cut and splice*, and *path crossover* (see, e.g., Ahuja et al. 2000).
- *Mutation*. This is another major natural selection operator, which spontaneously modifies the alleles of parent chromosomes to ensure both diversity (through exploration of new solutions) and quality

[16]The penalty values create significant gaps between the fitness values of feasible and infeasible solutions, which eventually lead to the elimination of the infeasible solutions in the next iterations of the algorithm.

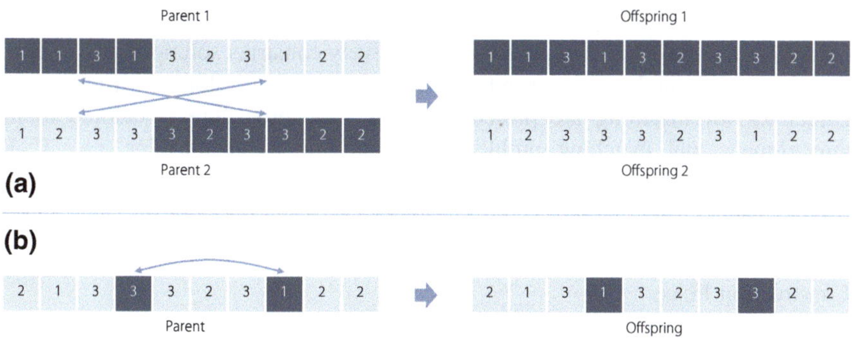

Fig. 5.7 Reproduction operators for the example of teaming for construction (Fig. 5.4). **a** Two-point crossover: parents 1 and 2 with fitness values of 722 and 676 (both infeasible) breed offsprings 1 (feasible) and 2 (infeasible) with fitness values of 714 and 706, respectively. **b** Swap mutation: parent with fitness value of 714 (feasible) breeds an offspring with fitness value of 722 (feasible)

(through exploitation of the parents' good traits) in the generated offsprings. Mutation operators are diverse. A common example is *swap mutation*, which randomly selects two genes and swaps their alleles (Fig. 5.7b). Examples of other popular mutation operators are *bit string* and *flip bit* mutation (see Mitchell 1996).

Step 5 *Optimization*. Steps 3 and 4 must be repeated until the algorithm converges. The main parameters that must be defined for efficient execution of the GA are the population size, the stopping criterion/criteria (e.g., no recent improvement or reaching a predefined number of iterations), and rates of crossover and mutation (i.e., what percentage of the population is modified using each operator).

5.3.2 *Greedy Randomized Adaptive Search Procedure (GRASP)*

GRASP is a constructive metaheuristic that progressively (1) constructs *greedy randomized* solutions, and (2) improves them through neighborhood search (Feo and Resende 1995). The first phase is accomplished by storing elements in a restricted candidate list (RCL) using a greedy function, and adding them to the solution, one at a time, according to their respective ranks in the RCL. In the second phase, the greedy randomized solution is improved via certain neighborhood search procedures. This procedure is repeated until the algorithm converges. GRASP is an efficient metaheuristic for best matching, especially for solving complicated instances such as biquadratic (Mavridou et al. 1998) and three-dimensional (Aiex et al. 2005) matching problems.

Logic

GRASP is composed of two phases that iteratively construct solutions and improve them through neighborhood search. Details of each phase along with application guidelines for matching are presented below. The procedure is tailored to the three-dimensional matching problem with one-to-one relationships, i.e., $1:1:1//+, OS$ [see Model (M 3.3)], for illustration. It, however, can be extended in a similar manner, and applied to other instances of matching with different dimensions, pairwise relationships, conditions, and/or criteria.

Construction Phase

Step 1 *Parameter setting*. Randomly select parameter $\alpha \in [0,1]$, create candidate list $CL = \{(i,j,k) \in I \times J \times K\}$, and set the best objective function value to $F^{\text{best}} = 0$.

Step 2 *Greedy randomized search*. Identify the maximum and minimum preference scores for $(i,j,k) \in CL$,

$$\bar{P} = \max\{P_{ijk}\colon (i,j,k) \in CL\},$$
$$\underline{P} = \min\{P_{ijk}\colon (i,j,k) \in CL\}.$$

Create $RCL = \{(i,j,k) \in CL : P_{ijk} \geq \bar{P} - \alpha(\bar{P} - \underline{P})\}$, and randomly select $(\hat{i},\hat{j},\hat{k}) \in RCL$.[17]

Step 3 *Update for feasibility.*[18] Update $CL \leftarrow CL \setminus \{(i,j,k) \in CL : i = \hat{i}, j = \hat{j}, \text{or } k = \hat{k}\}$

Step 4 *Construction*. If all the individuals are matched (i.e., $CL = \emptyset$), name the obtained solution S, update $F^{\text{best}} \leftarrow \max\{F^S, F^{\text{best}}\}$, and proceed to Step 5. Otherwise, return to Step 2.

Neighborhood Search Phase

Step 5 *Neighborhood definition*. Define x-exchange neighborhood for solutions S as

$$\Delta_x(S) = \{S'\colon \delta(S,S') = x\},$$

where $\delta(S,S')$ denotes the distance between solutions S and S', i.e.,

[17]Note that this RCL formula is valid for a maximization problem [Model (M 3.3)]. The formula for a minimization problem can be modified in a similar manner as follows: $RCL = \{(i,j,k) \in CL : P_{ijk} \leq \underline{P} + \alpha(\bar{P} - \underline{P})\}$.

[18]Note that this step satisfies the constraints (cardinality constraints in this case), and thus may be different for other instances of best matching, depending on their sets of constraints (e.g., capacity constraints).

Table 5.8 Movement times of boxes by robots to different locations (min)

	B_1			B_2			B_3		
	L_1	L_2	L_3	L_1	L_2	L_3	L_1	L_2	L_3
R_1	3.5	4.5	4.0	4.5	6.0	5.5	4.0	5.5	4.0
R_2	3.0	5.5	2.5	4.0	4.5	3.5	3.5	4.5	3.0
R_3	2.0	4.0	2.0	3.0	4.0	3.5	2.5	3.5	3.0

$$\delta(S, S') = \left(\sum_{i\in I}\sum_{j\in J}\sum_{k\in K}\left|\chi_{ijk}^{S} - \chi_{ijk}^{S'}\right|\right)/2.$$

Step 6 *Search*. Calculate the objective values of all solutions in neighborhood $\Delta_x(S)$. If a solution S' is detected with $F^{S'} > F^{\text{best}}$, update $F^{\text{best}} \leftarrow F^{S'}$, $S \leftarrow S'$, and return to Step 5. Otherwise, proceed to Step 7.

Step 7 *Optimization*. Repeat Steps 2 to 6 until the algorithm converges.

Example

Recall the example of material handling and warehouse robotics (Fig. 5.2), assuming each box is allowed to be located in any of the available locations. Table 5.8 summarizes the approximate movement times associated with each combination of robots, boxes, and locations. This is indeed a three-sided one-to-one best matching problem between sets $I = \{R_1, R_2, R_3\}$, $J = \{B_1, B_2, B_3\}$, and $K = \{L_1, L_2, L_3\}$ with the objective of minimizing the total movement time, i.e., $\sum_{i\in I}\sum_{j\in J}\sum_{k\in K} P_{ijk}\chi_{ijk}$, where

$$\chi_{ijk} = \begin{cases} 1, & \text{if robot } i \text{ carries box } j \text{ to location } k \\ 0, & \text{otherwise} \end{cases}, \quad \forall i \in I, j \in J, k \in K.$$

The GRASP described earlier solves the problem as follows.

Step 1 $\alpha = 0.2$ is selected at random. The initial candidate list includes all possible triplets, i.e., $CL = \{(R_1, B_1, L_1), \ldots, (R_3, B_3, L_3)\}$, and $F^{\text{best}} = 0$.

Step 2 From Table 5.8, the maximum and minimum movement times for $(i, j, k) \in CL$ are $\bar{P} = 6$ and $\underline{P} = 2$, respectively. Since the objective is to minimize the total movement times, $RCL = \{(i, j, k) \in CL : P_{ijk} \leq 2 + 0.2 \times (6 - 2) = 2.8\} = \{(R_3, B_1, L_1), (R_2, B_1, L_3), (R_3, B_1, L_3), (R_3, B_3, L_1)\}$. Triplet (R_2, B_1, L_3) is selected at random as the initial element of the solution (i.e., $\chi_{R_2, B_1, L_3} = 1$).

Step 3 Since (R_2, B_1, L_3) is selected as an element of the solution, all triplets with R_2, B_1, or L_3, i.e., (R_2, j, k), (i, B_1, k), (i, j, L_3), $\forall i \in I, j \in J, k \in K$, are eliminated from CL (to ensure feasibility).

Step 4 Steps 2 and 3 are repeated until all robots, boxes, and locations are matched. The following is found as the initial solution[19] S: R_2, B_1, L_3; R_1, B_2, L_2; R_3, B_3, L_1. The objective value for this solution is $F^S = F^{\text{best}} = 11$ minutes.

Step 5 2-exchange neighborhood[20] $\Delta_2(S)$ is selected, which includes 9 solutions that differ from S only in 2 matching decisions (e.g., S': R_3, B_1, L_3; R_1, B_2, L_2; R_2, B_3, L_1).

Step 6 Searching through neighborhood $\Delta_2(S)$, solution S': R_2, B_1, L_3; R_3, B_2, L_2; R_1, B_3, L_1 is obtained with the objective value of $F^{S'} = 10.5 < F^S$. Hence, S is replaced with S', $F^{\text{best}} \leftarrow 10.5$, and a new neighborhood search is performed on the new S by repeating Step 5. This process is continued until no improvement is detected, and then the algorithm proceeds to Step 7.

Step 7 If a predefined set of convergence criteria is satisfied, the algorithm stops and outputs the (near-)optimal solution. Otherwise, it returns to Step 2 for a new randomized search. In this small-sized example, the new S is indeed the optimal solution (although it is not unique), and thus $F^{\text{best}} = 10.5$ minutes is the minimum total movement of the robots given the best match of S^*: R_2, B_1, L_3; R_3, B_2, L_2; R_1, B_3, L_1.

5.3.3 Ant Colony Optimization *(*ACO*)*

ACO is a constructive metaheuristic inspired by the *foraging behavior* of ants in their search for food and the shortest path back to their nest (Dorigo et al. 1996). ACO is based on primitive behaviors of individual ants, which lead to highly intelligent behavior at the scale of colony or swarm, through efficient *interaction*. As finely quoted by Sociobiologist Edward O. Wilson:

> In theory at least, individual members of the swarm can profit from the discoveries and previous experience of all other members of the swarm during the search for food. This advantage can become decisive, outweighing the disadvantages of competition for food items, whenever the resource is unpredictably distributed in patches.

In ant colonies, this *autocatalytic* behavior is initiated after the first ant finds a food source. She then evaluates the quality and quantity of the food, and if adequate, carries a portion of it back to the nest, while depositing a *pheromone trail*. The pheromone functions as an indirect interaction mechanism between the ants, enabling a *sign-based stigmergy* that signals both the food source and a suggested path. Through a probabilistic behavior, the next ants either *exploit* the knowledge provided by the previous ants by following the strongest pheromone trail, or

[19] Note that due to the inherent randomness of GRASP, the obtained solutions are not necessarily unique.

[20] The size of 2-exchange neighborhood is $\binom{n}{2}^2$, where $n = |I| = |J| = |K|$.

explore new paths. Since pheromone is accumulated faster on shorter paths, however, this behavior of ants enables a *reinforcement learning mechanism* that eventually leads to the detection of the shortest path. ACO is an efficient metaheuristic for solving various hard best matching problems. A systematic description of the algorithm is presented next, together with examples and illustrations.

Logic

Although ACO is in the class of constructive metaheuristics, it can also be regarded as a population-based metaheuristic. In each iteration, A ants are generated to construct their own path (i.e., solution set) by creating its elements, one at a time, through biased exploration of new elements or exploitation of the knowledge provided by the previous ants. At the end, the shortest path (i.e., the best solution set) is rewarded by increasing its pheromone level. An ACO algorithm is presented below for solving the generalized best matching problem with interdependent preferences (IP), i.e., $M: 1/IP/+, OS$, which was discussed in Chaps. 2 (see Fig. 2.4) and 3 [see Model (M 3.9)].[21]

Step 1 *Initialization*. Generate A ants. Place each ant a, at random, at a current position $\pi_a = (i,j)$, which corresponds to a decision variable $\chi_{ij} = 1$. For each ant, define working memory $\omega_a = \{\pi_a\}$, tabu list $\tau_a = \emptyset$, and auxiliary parameters $\gamma_a \in [0,1]$ and $\lambda_a \in [0,1]$.

Step 2 *Tabu list update*. Given the current position $\pi_a = (i,j)$, update the tabu list τ_a by adding pairs (p,q) to it, where $\chi_{pq} = 1$ leads to the infeasibility of[22]

- *Cardinality constraints*: Update $\tau_a \leftarrow \tau_a \cup \{(p,q)\}$, where $p = i$ and $q \in J \backslash \{j\}$.
- *Capacity constraints*: Update $M_j \leftarrow M_j - 1$, and then $\tau_a \leftarrow \tau_a \cup \{(p,q)\}$, where $p \in I \backslash \{i\}$ and $q \in \{u : u \in J, M_u = 0\}$.

Step 3 *Search*. Randomly generate $\gamma \in [0,1]$. Select the new position $\hat{\pi}_a = (\hat{i},\hat{j})$ of ant a $(a = 1,\ldots,A)$ as follows:

- *Exploitation* $(\gamma < \gamma_a)$:

$$\left(\hat{i},\hat{j}\right) = \arg \max_{(p,q) \notin \tau_a \cup \omega_a} \left\{ \varphi_{(i,j),(p,q)} \Delta_{(p,q)} \right\} \tag{5.17}$$

- *Exploration* $(\gamma \geq \gamma_a)$: Select $\left(\hat{i},\hat{j}\right)$ based on the selection probability

[21]This basic ACO algorithm is based on the initial work by Dorigo et al. (1996). Other types of best matching problems can be solved using this or other extensions of the ACO algorithm in a similar fashion.

[22]Note that this procedure is for the matching problem $M: 1/IP/+, OS$, and may differ in other instances. The tabu list update ensures feasibility of the generated solutions.

$$\Pr(\hat{i},\hat{j}) = \begin{cases} \frac{\varphi_{(i,j),(\hat{i},\hat{j})}\Delta_{(\hat{i},\hat{j})}}{\sum_{(p,q)\in I\times J}\varphi_{(i,j),(p,q)}\Delta_{(p,q)}}, & \text{if}(\hat{i},\hat{j}) \notin \tau_a \cup \omega_a, \\ 0, & \text{otherwise.} \end{cases} \quad (5.18)$$

$\varphi_{(i,j),(p,q)}$ denotes the pheromone level between pairs (i.e., positions) (i,j) and (p,q), and $\Delta_{(p,q)}$ denotes the amount of increase in the objective value if (p,q) is selected. Update the working memory $\omega_a \leftarrow \omega_a \cup \{(\hat{i},\hat{j})\}$, and current position $\pi_a \leftarrow \hat{\pi}_a$.

Step 4 *Local pheromone update.*[23] Update $\varphi_{(i,j),(\hat{i},\hat{j})} \leftarrow (1-\lambda_a)\varphi_{(i,j),(\hat{i},\hat{j})} + \lambda_a\varphi_0$, where φ_0 is a positive constant

Step 5 *Construction.* If all ants have complete paths (i.e., represent complete solution sets), proceed to Step 6. Otherwise, return to Step 2

Step 6 *Global pheromone update.*[24] Find ant a^* with the shortest path, and update the global pheromone level of her path $\varphi_{a^*}^{\text{path}} \leftarrow (1-\lambda_{a^*})\varphi_{a^*}^{\text{path}} + \lambda_{a^*}F_{a^*}$, where F_{a^*} denotes the objective value corresponding to the path (i.e., solution set) found by ant a^*

Step 7 *Optimization.* Repeat Steps 2 to 6 until the algorithm converges

Example

Recall the example of resident matching shown in Fig. 2.4, and assume that the housing community manager is trying to rematch residents $i_1,\ldots,i_7$ to houses j_1 and j_2, which have four and three bedrooms, respectively. This is indeed a generalized matching problem with IP, i.e., $M : 1/IP/+, OS$ [Model (M 3.9)], between sets $I = \{i_1,\ldots,i_7\}$ and $J = \{j_1,j_2\}$. The matching is based on the preferences of the residents for the houses (Table 5.9) as well as their mutual influences on each other's preferences (Table 5.10), based on their roommate matching inputs.

With this information, ACO can help the community manager in finding the best match between residents and houses as follows:

Step 1 Assume 3 ants are generated in the first iteration, where $\pi_1 = (i_2,j_1)$, $\pi_2 = (i_4,j_2)$, $\pi_3 = (i_3,j_2)$, and $\gamma_a = 0.5$, $\lambda_a = 0.5$, $\tau_a = \emptyset$, for $a = 1,2,3$. The working memories are therefore $\omega_1 = \{(i_2,j_1)\}$, $\omega_2 = \{(i_4,j_2)\}$, and $\omega_3 = \{(i_3,j_2)\}$.

Step 2 Given the initial positions and constraints of the problem (i.e., cardinality: each resident must be matched to exactly one house; capacity: four and three bedrooms available at j_1 and j_2, respectively), the tabu lists of the ants are updated as follows $\tau_1 = \{(i_2,j_1)\}$, $\tau_2 = \{(i_4,j_1)\}$, and $\tau_3 = \{(i_3,j_1)\}$.

[23] Local pheromone update imitates the *pheromone trail evaporation* in real ant systems.

[24] Global pheromone update imitates the property of *differential pheromone trail accumulation* in real ant systems—the reinforcement learning mechanism.

Table 5.9 Preferences of residents for houses (%)

	i_1	i_2	i_3	i_4	i_5	i_6	i_7
j_1	70	65	75	60	90	70	100
j_2	95	50	80	35	80	70	60

Table 5.10 Influences of residents on each other's preferences ($\alpha_{ii'} \in [-1, 1]$)

i	i'						
	i_1	i_2	i_3	i_4	i_5	i_6	i_7
i_1	–	−0.2	0.8	−0.7	0.5	0.9	−0.3
i_2	−0.1	–	−0.3	0.4	−0.4	0.1	−0.8
i_3	0.6	−0.5	–	−0.8	0.8	−0.9	−0.8
i_4	1.0	−0.4	−0.7	–	−0.2	0.7	−0.4
i_5	0.8	−0.9	0.6	−0.2	–	−0.5	−0.4
i_6	0.6	−0.6	0.6	−0.8	−0.2	–	−0.1
i_7	−0.5	−0.5	0.0	−0.8	0.3	0.1	–

Step 3 Since there is no prior knowledge in the first iteration, the pheromone levels of all paths are assumed constant and equal, and thus, path selection is merely performed through biased exploration. Assume $\hat{\pi}_1 = (i_5, j_1)$, $\hat{\pi}_2 = (i_1, j_2)$, and $\hat{\pi}_3 = (i_7, j_1)$ are selected by the ants as their second positions, which are then updated as their current positions. The working memories are also updated as $\omega_1 = \{(i_2, j_1), (i_5, j_1)\}$, $\omega_2 = \{(i_4, j_2), (i_1, j_2)\}$, and $\omega_3 = \{(i_3, j_2), (i_7, j_1)\}$.

Step 4 The local pheromones $\varphi_{(i_2,j_1),(i_5,j_1)}$, $\varphi_{(i_4,j_2),(i_1,j_2)}$, and $\varphi_{(i_3,j_2),(i_7,j_1)}$ must then be updated for ants 1, 2, and 3, respectively.[25]

Step 5 Repeating Steps 2–4 results in a complete and feasible path (i.e., solution set) for each ant. Consider the following solutions obtained in the first iteration (Details of the calculations are omitted for the sake of brevity):

	Path (ω_a)	Objective value
$a = 1$	$\{(i_2, j_1), (i_5, j_1), (i_6, j_1), (i_1, j_2), (i_3, j_2), (i_7, j_1), (i_4, j_2)\}$	232.5
$a = 2$	$\{(i_4, j_2), (i_1, j_2), (i_6, j_1), (i_2, j_1), (i_3, j_1), (i_5, j_2), (i_7, j_1)\}$	300.0
$a = 3$	$\{(i_3, j_2), (i_7, j_1), (i_1, j_2), (i_5, j_1), (i_2, j_1), (i_6, j_2), (i_4, j_1)\}$	434.5

Step 6 The third ant found a significantly better route compared to the other two ants. Thus, the pheromone level of the entire path is increased proportional to its objective value.

[25]Note that parameter φ_0 must be selected such that the local pheromone levels decrease, in order to imitate the pheromone evaporation in real ant systems.

Step 7 Repeating Steps 2–6 leads the algorithm towards the optimal solution. Details of the calculations are omitted for the sake of brevity. The optimal matching of residents to the houses is obtained in a few iterations as follows (✓: Matched):

	i_1	i_2	i_3	i_4	i_5	i_6	i_7
j_1	✓		✓		✓	✓	
j_2		✓		✓			✓
Z^*	666.5 (average resident satisfaction of $666.5/7 = 95.2\,\%$)						

5.3.4 *Tabu Search*

This optimization technique is a *memory-based* local search metaheuristic (Glover 1986). The neighborhood search follows certain prohibition strategies that mark previously visited solutions (marking them either temporarily or permanently) as forbidden or *tabu*,[26] in order to prevent cycling and improve the efficiency of search mechanism (Glover 1986). The algorithm starts with an initial solution S and continually *explores* through its neighborhood while *exploiting* the knowledge of tabu points, until a set of convergence criteria is satisfied. Similar to the other metaheuristics discussed so far, tabu search can be used for solving an extensive range of best matching problems.

Logic

A unique property of tabu search compared to other neighborhood search heuristics is the use of short-/long-term memories for enhancing the quality and efficiency of search by eliminating undesirable alternatives from the list of candidate solutions. That is, after a component of the solution set S is eliminated in the solution set S', it becomes tabu and therefore, is not selected for a certain number of iterations. This gap is referred to as *tabu time*. It may be static or dynamic, and depends on various factors, such as the magnitude of change in the solution quality after removing that component. A generic tabu search algorithm for best matching is presented below,[27] which can be modified and applied for solving various types of best matching problems.

[26]The term "tabu" originates from Tongan *tapu*, referring to "prohibited" or "forbidden" things (Merriam-Webster dictionaries).

[27]There is a wide variety of extensions for tabu search; however, the basic idea for all is almost the same as the one presented here. Interested readers are encouraged to read the user's guide for tabu search by Glover et al. (1993).

Step 1 *Initialization*. The starting point for the search is an initial solution S, which can be obtained in various ways. The quality of the initial solution has a significant influence on the quality of search, and thus, random generation is not a preferred approach. The recommended approaches for best matching applications include (but are not limited to) the heuristics discussed in Sect. 5.2 (e.g., the Lagrangian relaxation method), and the construction phase of GRASP (Sect. 5.3.2, Logic: Steps 1–4).

Step 2 *Neighborhood search*. This step includes delineation of the neighborhood for S, as well as the search procedure. This step is similar to the neighborhood search phase of GRASP (Sect. 5.3.2, Logic: Steps 5–6). x-exchange neighborhoods $\Delta_x(S)$ such as *shift* ($x = 1$) and *swap* ($x = 2$) are common, together with more advanced neighborhoods such as *ejection chains* and *path relinking* (see, e.g., Yagiura et al. 2006).

Step 3 *Tabu time updates*. After a variable $\hat{\chi}$ leaves the solution, i.e., $\hat{\chi} = \{\chi : \chi \in S, \chi \notin S'\}$, it enters the tabu list, and therefore its tabu time (i.e., the number of iterations it has to spend in the tabu list) must be calculated. Naturally, the tabu time $t(\hat{\chi})$ must be proportional to the cost/benefit of removing $\hat{\chi}$ from the solution set, i.e.,

$$t(\hat{\chi}) \propto \frac{c(\hat{\chi})}{b(\hat{\chi})},$$

or

$$t(\hat{\chi}) = \varepsilon \frac{c(\hat{\chi})}{b(\hat{\chi})}, \quad \forall \hat{\chi} = \{\chi : \chi \in S, \chi \notin S'\}, \tag{5.19}$$

where $c(\hat{\chi})$ and $b(\hat{\chi})$ denote the associated cost and benefit of removing $\hat{\chi}$, respectively,[28] and ε is a positive constant. After each iteration, the tabu times are reduced ($t(\chi) \leftarrow t'(\chi)$, where $t'(\chi) < t(\chi)$), and variables χ are freed from the tabu list after $t(\chi) = 0$.

Step 4 *Optimization*. Steps 2 and 3 must be repeated until the algorithm converges.

Example

Recall the example of operator, machine, and job matching from Chap. 2 (see Fig. 2.2c). The shop is composed of three machines that are identical (i.e., can process all the jobs) and can be run by any of the operators. Due to the differences in their level of experience and expertise, however, each operator has different performance on each machine in terms of processing time. In addition, although the machines are identical, the final quality of each job varies, depending

[28]The cost and benefit parameters discussed here represent the integrated impact (cost: negative; benefit: positive) of eliminating $\hat{\chi}$ from the solution, on the objective function value, resource utilization, and other relevant factors.

Table 5.11 Final quality of jobs for different pairs of machine and operator (normalized in [0, 1])

	M_1			M_2			M_3		
	O_1	O_2	O_3	O_1	O_2	O_3	O_1	O_2	O_3
J_1	0.95	0.63	0.75	0.86	0.65	0.85	0.88	0.57	0.76
J_2	0.80	0.70	0.72	0.90	0.77	0.89	0.97	0.55	0.75
J_3	0.92	0.72	0.84	0.88	0.65	0.83	0.97	0.58	0.80
J_4	0.84	0.74	0.82	0.80	0.79	0.75	0.97	0.59	0.74
J_5	0.81	0.72	0.81	0.92	0.71	0.79	0.85	0.63	0.71

Table 5.12 Processing times for different combinations of machine, job, and operator (h)

	M_1			M_2			M_3		
	O_1	O_2	O_3	O_1	O_2	O_3	O_1	O_2	O_3
J_1	2.93	4.01	3.22	3.72	5.00	2.92	3.73	3.88	4.58
J_2	3.64	4.81	3.97	3.03	3.59	3.48	3.03	4.62	4.18
J_3	2.88	5.60	3.15	3.11	4.90	3.47	3.07	5.43	4.11
J_4	2.92	3.86	4.24	3.51	4.09	4.26	3.61	4.97	4.21
J_5	3.19	5.31	3.39	3.18	3.84	3.32	3.40	4.71	3.95

on its allocated machine. Tables 5.11 and 5.12 summarize the process information for three machines (set $I = \{M_1, \ldots, M_3\}$), five jobs (set $J = \{J_1, \ldots, J_5\}$), and three operators (set $K = \{O_1, \ldots, O_3\}$). The goal of the shop manager is to find the best match between the machines, jobs, and operators such that the jobs are completed in one shift (8 h) with the maximum quality possible.

This is a three-sided resource constrained best matching problem, which can be formulated as follows:

$$\begin{aligned}
\text{Maximize} \quad & Z = \sum_{i \in I} \sum_{j \in J} \sum_{k \in K} P_{ijk} \chi_{ijk} \\
\text{Subject to} \quad & \sum_{i \in I} \sum_{k \in K} \chi_{ijk} = 1, \quad \forall j \in J, \\
& \sum_{j \in J} \sum_{k \in K} r_{ijk} \chi_{ijk} \leq 8, \quad \forall i \in I, \\
& \chi_{ijk} \in \{0, 1\}, \quad \forall i \in I, j \in J, k \in K.
\end{aligned}$$

In the model above, P_{ijk} denotes the final quality of product j, if processed on machine i and by operator k. The first set of constraints ensures that each job is matched to exactly one *unique* pair of machine and operator (i.e., one-to-one matching of machines and operators), and the second set of constraints ensures that the total processing time of each machine does not violate the length of the shift (i.e., 8 h). Tabu search can be applied for solving this problem as follows:

Step 1 The construction phase of GRASP can be applied for initialization,[29] where

$$\bar{P} = \max \left\{ P_{ijk}/r_{ijk} : (i,j,k) \in CL \right\},$$
$$\underline{P} = \min \left\{ P_{ijk}/r_{ijk} : (i,j,k) \in CL \right\}.$$

The other notable difference here compared to the example presented in Sect. 5.3.2 is that this is a generalized matching problem involving capacity constraints. Hence, the update for feasibility step (Step 3 of GRASP) must be performed as follows: After variable $\chi_{\hat{i}\hat{j}\hat{k}}$ enters the solution set, variables χ_{ijk} must be removed from the set of candidate lists where

- $i = \hat{i}, j \in J, k \in K \backslash \{\hat{k}\}$,
- $i \in I \backslash \{\hat{i}\}, j \in J, k = \hat{k}$, or
- $i \in I \backslash \{\hat{i}\}, j = \hat{j}, k \in K \backslash \{\hat{k}\}$.

The following solution is an initial solution for this problem obtained using the construction phase of GRASP: $\chi_{M_1,J_3,O_1} = \chi_{M_1,J_4,O_1} = \chi_{M_2,J_1,O_3} = \chi_{M_2,J_2,O_3} = \chi_{M_3,J_5,O_2} = 1$, and $\chi_{ijk} = 0$ for other $i \in I, j \in J, k \in K$, with the objective value of $Z = 4.13$.

Step 2 Various types of neighborhoods can be applied; e.g., 2-exchange neighborhood similar to the robotic material handling example discussed in Sect. 5.3.2.

Step 3 The tabu times for the leaving variables can be defined according to their associated quality levels and processing times; i.e., $c(\chi_{ijk}) \equiv r_{ijk}$ and $b(\chi_{ijk}) \equiv P_{ijk}$ [see Eq. (5.19)].

Step 4 Repeating Steps 2–3 yields the following optimal solution after a reasonably small number of iterations (Details are omitted for the sake of brevity):

	M_1			M_2			M_3		
	O_1	O_2	O_3	O_1	O_2	O_3	O_1	O_2	O_3
J_1						✓			
J_2						✓			
J_3							✓		
J_4							✓		
J_5		✓							
Z^*	4.40 (Total final quality of jobs based on the normalized inputs in Table 5.11).								

[29]Note that initialization can also be done randomly, which is likely to result in infeasible solutions. In general, several instances of tabu search allow infeasible solutions and moves in the neighborhood, with the goal of finding better feasible solutions at the end. In those cases, infeasible solutions must be penalized.

5.4 Summary

This chapter presented a portfolio of the best algorithms for best matching (Table 5.13). The Hungarian method was introduced as a computationally efficient algorithm for solving one-to-one matching problems.[30] The discussion was followed by the deferred acceptance algorithm, an efficient procedure for stable (but not necessarily optimal) market design.[31] These methods (and their variants) provide promising mechanisms for solving two-sided matching problems with one-to-one relationships. For generalized and higher dimension matching problems, however, more advanced algorithms are required, in order to handle their inherent computational complexity. The Lagrangian relaxation method was introduced as an example of approximation heuristics for obtaining good upper/lower bounds in reasonable time. *Set partitioning* is another approximation method based on column generation heuristics, which—similar to the Lagrangian relaxation method—transforms a generalized matching problem into $|J|$ independent knapsack problems for generating upper/lower bounds (see Cattryse and van Wassenhove 1992). *Linear programming relaxation* is another example of approximation methods, which relaxes the integrality constraints by replacing the binary matching variables with nonnegative real variables, and recasts the generalized matching problem as a special generalized network problem solvable in polynomial time (Trick 1992). The bounds generated by the aforementioned (and similar) methods can be used for solving various matching problem, either by extensions of the same methods, or other methods such as branch-and-bound.

The introduced approximation methods can also be used for performance evaluation and validation of metaheuristics, which naturally provide sufficiently good but not necessarily optimal solutions to more advanced and complicated matching problems that are not solvable with the basic methods discussed. GA, GRASP, ACO, and tabu search were introduced and illustrated as popular metaheuristic algorithms for solving more complicated and unstructured best matching problems. Examples of other well-known metaheuristics for best matching that are not discussed here are simulated annealing (see, e.g., Wilhelm and Ward 1987) and artificial neural networks (see, e.g., Funabiki 1992). These algorithms and their extensions can be customized and applied for solving any special instances of best matching problems under different conditions (e.g., interdependencies between preferences, resource sharing, and layered matching) and criteria. It will be shown in Chap. 6 that these algorithms can also be used for modeling and analysis of the evolution of distributed systems by incorporating the D+ dimension of the PRISM taxonomy of best matching in the decision-making processes.

[30] It can be shown that the best time complexity for the Hungarian method is $O(n^3)$.

[31] The time complexity of the deferred acceptance algorithm is $O(n^2)$.

Table 5.13 Recommended algorithms for best matching based on the PRISM taxonomy[a]

	HM[b]	DAA	LRM	BBM	GA	GRASP	ACO	TS
$1:1//+,OS$	✓	✓						
$M:1//+,OS$			✓	✓	✓	✓	✓	✓
$1:1:1//+,OS$				✓	✓	✓	✓	✓
$M:1/RC/+,OS$			✓	✓	✓	✓	✓	✓
$M:1/RC,PR/+,OS$				✓	✓	✓	✓	✓
$M:1/RC,RS/+,OS$					✓	✓	✓	✓
$1:1/IP/+,OS$				✓	✓	✓	✓	✓
$M:1/IP/+,OS$				✓	✓	✓	✓	✓
$1:1/LM/+,OS$				✓	✓	✓	✓	✓
$1:1//+,OS/DI,ES$					✓		✓	

[a]Note that each recommended (or similar) algorithm must be modified for solving each of the listed (and similar) best matching problems

[b]*HM* Hungarian method, *DAA* deferred acceptance algorithm, *LRM* Lagrangian relaxation method, *BBM* branch-and-bound method, *GA* genetic algorithm, *GRASP* greedy randomized adaptive search procedure, *ACO* ant colony optimization, *TS* Tabu search, ✓: recommended algorithm

Problems

5.1. *Compare and contrast*: Hungarian method versus deferred acceptance algorithm.

(a) Use the Hungarian method to solve the problem of matching doctors and hospitals discussed in Sect. 5.2.2 (Fig. 5.3).
(b) Use the deferred acceptance algorithm to solve the problem of matching robots and boxes discussed in Sect. 5.2.1 (Fig. 5.2).
(c) Do the Hungarian method and the deferred acceptance algorithm yield similar or different results in the two cases above? Why? Explain your answer.

5.2. Recall the example of intelligent warehouse management system discussed in Sect. 5.2.1 (Fig. 5.2). Assume that the locations of the boxes are not predefined and must be decided. That is, the problem is to find the best match between the sets of robots $I = \{R_1, R_2, R_3\}$, boxes $J = \{B_1, B_2, B_3\}$, and locations $K = \{L_1, L_2, L_3\}$, such that the total movement time of the robots is minimized. The movement time data are presented in Table 5.14.

Table 5.14 Movement times of boxes by robots (min)

	L_1			L_2			L_3		
	B_1	B_2	B_3	B_1	B_2	B_3	B_1	B_2	B_3
R_1	3.5	2.8	2.4	4.9	5.0	5.2	4.6	3.3	4.0
R_2	3.0	4.5	2.0	4.0	4.5	3.3	3.7	4.2	2.5
R_3	2.0	1.5	2.6	2.3	4.0	4.1	2.4	3.2	3.0

(a) Use the Hungarian method to find the best match between sets J (boxes) and K (locations), considering the mean values of the movement times associated with the three robots. Use the Hungarian method once again to match the outcome of J and K matching with set I (robots).
(b) Use the Hungarian method to find the best match between sets I (robots) and J (boxes), considering the mean values of the movement times associated with the three location points. Use the Hungarian method once again to match the outcome of I and J matching with set K (locations).
(c) Use the Hungarian method to find the best match between sets I (robots) and K (locations), considering the mean values of the movement times associated with the three boxes. Use the Hungarian method once again to match the outcome of I and K matching with set J (boxes).
(d) Compare the results of (a), (b), and (c). Are they similar or different? Why?
(e) Can you apply the Hungarian method for three-sided matching? Explain and illustrate your answer.

5.3. Recall the example of teaming for construction discussed in Sect. 5.2.3 (Fig. 5.4).

(a) Use the deferred acceptance algorithm to match employees and construction projects, assuming that the employees propose to the projects (project managers).
(b) Repeat (a) assuming that the projects propose to the employees.
(c) Are the results of (a) and (b) the same or different? Explain why.
(d) What changes did you have to make in the original deferred acceptance algorithm in order to solve this one-to-many matching problem? (Remember that the algorithm presented in Sect. 5.2.2 is designed for one-to-one matching.)

5.4. Consider a manufacturing shop with two unrelated machines M_1 and M_2. The shop has received an order to process seven different parts P_1 to P_7. The available operational times of machines M_1 and M_2 for this order are 10 and 20 h, respectively. The production cost and processing time data are summarized in Tables 5.15 and 5.16, respectively.

(a) Formulate the above problem as a generalized resource-constrained matching problem minimizing total production cost; i.e., $M : 1/RC/-, OS$.
(b) Use the Lagrangian relaxation methods (L1 and L2) to solve this matching problem.

Table 5.15 Production costs ($)

Machines	Parts						
	P_1	P_2	P_3	P_4	P_5	P_6	P_7
M_1	8	7	4	5	4	9	1
M_2	9	10	6	3	6	4	2

Table 5.16 Processing times (h)

Machines	Parts						
	P_1	P_2	P_3	P_4	P_5	P_6	P_7
M_1	1	4	8	2	4	2	1
M_2	8	3	7	8	9	8	1

(c) Use the branch-and-bound method to solve this matching problem.
(d) Compare the results of (b) and (c). Do you notice any difference? Explain.
(e) Comment on the advantages and limitations of the Lagrangian relaxation and branch-and-bound methods in solving various instances of matching problems.

5.5. Construct a GA (Genetic Algorithm) for the three-sided matching instance presented in Problem 5.2 (i.e., matching boxes, robots, and locations).

(a) What encoding scheme do you suggest? Does it ensure feasibility? Illustrate.
(b) Create a heuristic approach for generating the initial population. How do you ensure acceptable *quality* and *feasibility* of the generated chromosomes? Illustrate.
(c) Compare your proposed initialization procedure in (b) with GRASP.
(d) Suggest an efficient reproduction procedure including crossover and mutation operators.
(e) Besides the initialization and reproduction procedures, what other factors may influence the performance of your GA? Explain.

5.6. Recall the example of resident matching discussed in Sect. 5.3.3.

(a) Multiply all the α values shown in Table 5.10 by 0.5, and run the ACO (Ant Colony Optimization) algorithm presented in Sect. 5.3.3 again to find the optimal match.
(b) Repeat (a) by setting all α values equal to zero.
(c) Plot the optimal objective function values obtained in (b), (a), and the example presented in Sect. 5.3.3 (with the original α values), in that order. What do you observe? How do you justify the impact of IP (Interdependent Preferences) on the outcomes of matching using these results? Explain.

5.7. *Compare and contrast*: GA *versus* tabu search.

(a) Develop a GA to solve the operator, machine, and job matching problem discussed in Sect. 5.3.4.
(b) Develop a tabu search to solve the box, robot, and location matching problem discussed in Problems 5.2 and 5.5.
(c) Which algorithm (GA or tabu search) do you prefer considering their performances in the above two cases? Explain and illustrate your answer.

5.8. Define the matching process as m and the best matching process as m^*. Define a feasible, tolerated match (i.e., all items within a matched set are within design tolerance limits) as M. Define the best match as M^*.

(a) Can m yield M^*? Illustrate your answer with algorithms described in this chapter.
(b) Does applying m^* guarantee that it will yield M^*? Illustrate and explain your answer.
(c) Select two alternative algorithms (i.e., m_1 and m_2) that can solve the same best matching model. Would M_1^* be the same as M_2^*? Explain why or why not.
(d) Following (c), how can you compare m_1 and m_2 in terms of their matching process efficiency?

References

Ahuja, R. K., Orlin, J. B., & Tiwari, A. (2000). A greedy genetic algorithm for the quadratic assignment problem. *Computers & Operations Research, 27*, 917–934.

Aiex, R. M., Resende, M. G. C., Pardalos, P. M., & Toraldo, G. (2005). GRASP with path relinking for three-index assignment. *INFORMS Journal on Computing, 17*, 224–247.

Burkard, R., Dell'Amico, M., & Martello, S. (2012). Assignment problems. SIAM.

Cattrysse, D. G., & van Wassenhove, L. N. (1992). A survey of algorithms for the generalized assignment problem. *European Journal of Operational Research, 60*, 260–272.

Dakin, R. J. (1965). A tree search algorithm for mixed integer programming problems. *Computer Journal, 8*, 250–255.

Dorigo, M., Maniezzo, V., & Colorni, A. (1996). Ant system: Optimization by a colony of cooperating agents. *IEEE Transactions on Systems, Man, and Cybernetics-Part B: Cybernetics, 26*, 29–41.

Feo, T. A., & Resende, M. G. C. (1995). Greedy randomized adaptive search procedures. *Journal of Global Optimization, 6*, 109–133.

Fisher, M. L. (1981). The Lagrangian relaxation method for solving integer-programming problems. *Management Science, 27*, 1–18.

Fisher, M. L. (1985). An applications oriented guide to Lagrangian relaxation. *Interfaces, 15*, 10–21.

Funabiki, N. (1992). A neural network parallel algorithm for channel assignment problems in cellular radio networks. *IEEE Transactions on Vehicular Technology, 41*, 430–437.

Gale, D., & Shapley, L. S. (1962). College admission and stability of marriage. *The American Mathematical Monthly, 69*, 9–15.

Glover, F. W. (1986). Future paths for integer programming and links to artificial intelligence. *Computers & Operations Research, 13*, 533–549.

Glover, F. W., Taillard, E., & Taillard, E. (1993). A user's guide to tabu search. *Annals of Operations Research, 41*, 1–28.

Held, M., & Karp, R. M. (1970). The traveling-salesman problem and minimum spanning trees: Part I. *Operations Research, 18*, 1138–1162.

Held, M., & Karp, R. M. (1971). The traveling-salesman problem and minimum spanning trees: Part II. *Mathematical Programming, 1*, 6–25.

Jeet, V., & Kutanoglu, E. (2007). Lagrangian relaxation guided problem space search heuristic for generalized assignment problems. *European Journal of Operational Research, 182*, 1039–1056.

Kuhn, H. W. (1955). The Hungarian method for the assignment problem. *Naval Research Logistics Quarterly, 2*, 83–97.

Land, A.H., & Doig, A.G. (1960). An automatic method of solving discrete programming problems. *Econometrica, 28*, 497–520.

Lovasz, L., & Plummer, M.D. (1991). *Matching theory*. Elsevier Science Publishers, B.V.

Luke, S. (2014). *Essentials of metaheuristics* (2nd ed.). Department of Computer Science, George Mason University.

Martello, S., Pisinger, D., & Toth, P. (1999). Dynamic programming and strong bounds for the 0-1 knapsack problem. *Management Science, 45*, 414–424.

Martello, S., Pisinger, D., & Toth, P. (2000). New trends in exact algorithms for the 0-1 knapsack problem. *European Journal of Operational Research, 123*, 325–332.

Mavridou, T., Pardalos, P. M., Pitsoulis, L. S., & Resende, M. G. C. (1998). A GRASP for the biquadratic assignment problem. *European Journal of Operational Research, 105*, 613–621.

Mitchell, M. (1996). *An introduction to genetic algorithms*. Cambridge: MIT Press.

Öncan, T. (2007). A survey of the generalized assignment problem and its applications. *INFOR, 45*, 123–141.

Ross, G. T., & Soland, R. M. (1975). A branch and bound algorithm for the generalized assignment problem. *Mathematical Programming, 8*, 91–103.

Roth, A. E. (2008). Deferred acceptance algorithms: history, theory, practice, and open questions. *International Journal of Game Theory, 36*, 537–569.

Trick, M. A. (1992). Linear relaxation heuristic for the generalized assignment problem. *Naval Research Logistics, 39*, 137–151.

Wilhelm, M. R., & Ward, T. L. (1987). Solving quadratic assignment problems by 'simulated annealing'. *IIE Transactions, 19*, 107–119.

Yagiura, M., Ibaraki, T., & Glover, F. W. (2006). A path relinking approach with ejection chains for the generalized assignment problem. *European Journal of Operational Research, 169*, 548–569.

Chapter 6
Dynamic and Distributed Matching

Abstract This chapter presents mechanisms for real-time optimization of dynamic and uncertain matching processes, and control of distributed decision-makers in the system. Several numerical examples are presented for illustrating the mechanics and applications of each method. The purpose is to demonstrate the fundamentals and major classes of real-time optimization techniques, show how to model distributed matching processes via multi-agent systems and task administration protocols, and discuss major challenges of such dynamic and distributed systems associated with "AI"—*artificial intelligence*; *analytics and informatics*.

6.1 Why Are Static and Centralized Algorithms not Always Sufficient?

The answer to this question is twofold:

(1) The inherent *dynamicity* and *uncertainty* of processes and the three main dimensions of best matching hinder the application of classic static algorithms in practice. The primary assumption of all the optimization algorithms presented in Chap. 5 is that the three dimensions of a best matching process (i.e., D1: Sets; D2: Conditions; D3: Criteria) are fixed. It is assumed in all those (and similar) algorithms that the number of sets, the number of individuals in each set, their pairwise relationships, the type and formulation of matching conditions, the number and type of criteria, and the mutual preferences of individuals are static and not subject to any changes over time. This assumption, however, may not be valid in many practical instances, as emphasized by the additional dimension of the PRISM taxonomy of best matching—*time or progression* (D+). The D+ dimension simply considers dynamic variations in the characteristics of the main three dimensions of a best

---

The original version of this chapter was revised: For detailed information please see erratum. The erratum to this chapter is available at 10.1007/978-3-319-46070-3_9

M. Moghaddam and S.Y. Nof, *Best Matching Theory & Applications*, Automation, Collaboration, & E-Services 3, DOI 10.1007/978-3-319-46070-3_6

matching process (see, e.g., Model (M 3.11) in Chap. 3). Below are several practical examples:

- *Selective assembly*. In the bolt and nut matching example in Chap. 1 (see Fig. 1.1), it was assumed that there are fixed and equal numbers of bolts and nuts. With that assumption, the one-to-one matching problem can be solved, with no trouble, by the Hungarian method (see Chap. 5). In a real assembly system where the parts (e.g., bolts and nuts) are dynamically arriving on two parallel conveyors/feeders, however, matching the dynamically changing elements of two (or more) sets of subassembly parts is not a simple task.
- *Scheduling*. In matching a set of jobs to a set of servers (e.g., Fig. 2.2c), it is typically assumed that the servers have constant performance rates and/or the precedence relationships of jobs are fixed. There are several approaches for solving these types of classic scheduling problems. In practice, however, it may not be known how many jobs must be processed, what the best sequence is, when the jobs are to be released, and/or what the processing times are. Such uncertainties introduce relative complexities to the matching process.
- *Storage allocation*. In an intelligent warehouse (e.g., Fig. 5.1), a core operational challenge is to define which material handling robot carries which box to which storage location. Different algorithms were suggested in Chap. 5 for solving a static version of this problem. Similar to the previous examples, however, practical instances of this problem may involve dynamicity of sets (e.g., busy/available robots; arriving streams of boxes), and/or process conditions (e.g., movement times vary with location of robots and size/destination of boxes). In addition, in *affinity-based* storage allocation,[1] the preferences of boxes for storage locations depend on the current inventory and configuration of the warehouse, which are indeed functions of time.
- *Networked manufacturing*. Design of CNO (Collaborative Networked Organizations) (see Fig. 2.9) involves matching distributed physical and virtual resources to distributed organizations and tasks, which may also be physical or virtual. In such complex systems with highly variable characteristics (e.g., dynamic coalitions of organizations; association/dissociation of resources or participants; variable task arrivals, requirements, sharing protocols), the classic static algorithms for matching may not be sufficient for finding the best real-time configuration of CNO.

The above (and similar) examples highlight the substantial need for more *adaptive* and *real-time optimization* (RTO) methods for best matching in practice.

---

[1]Modern storage allocation procedures suggest locating products with higher affinity (i.e., products that are ordered with each other more frequently) close to each other, in order to save material handling time. The affinity of products is calculated based on historical demand data and using data mining heuristics (see, e.g., Li et al. 2015).

(2) Besides being static, the optimization algorithms discussed in Chap. 5 are also *centralized*, which limits their application to cases where the entire matching process is designed by a single decision-maker (see Fig. 4.1). As discussed in Chap. 4, however, in many best matching instances, multiple decision-makers may be involved with various relative decision-making abilities (e.g., hierarchy; heterarchy) and relationships (e.g., competitive; collaborative). In such settings, the key for successful design of best matching processes is to develop effective *protocols* for *distributed control* with respect to the distribution of decisional abilities, as well as the nature of interactions among decision-makers.

This chapter addresses the above challenges by elaborating on the notion of RTO, its mechanisms and applications for best matching, and by introducing useful distributed control architectures and protocols for best matching in distributed systems. Basic *processes of matching* are introduced in dynamic settings, where one or multiple decision-makers continuously design and, if necessary, modify matches with regard to real-time changes.

6.2 Real-Time Optimization

Change is an inevitable phenomenon in almost every natural or artificial system. In natural systems, individuals *adapt* to environmental changes by means of natural selection, in order to enhance their fitness and survive. Although such adaptive traits are innate and *automatic* in natural systems, artificial systems are clearly not "born" with such level of intelligence, and require high-level design and control mechanisms in order to effectively adapt to potential changes in their characteristics and environment (see, e.g., the examples of assembly, scheduling, warehouse, and manufacturing systems discussed earlier). In a best matching process, specifically, "change" is formally defined as *a process in* $\Delta = t_1 - t_0$ $(t_1 > t_0)$, *where at least one characteristic of sets, conditions, or criteria in* t_1 *is defined differently from* t_0 (Fig. 6.1). The algorithms for design and optimization of dynamic best matching processes, therefore, must incorporate potential changes in their decision-making procedures.

In systems with ambiguous and dynamically changing domains and characteristics, in general, optimization approaches are essentially classified into three categories:

(1) *Robust optimization*. When there is no possibility for real-time measurement, feedback, and modification of the design, robust optimization is one of the best approaches to take. Robust optimization applies Wald's maximin model (Wald 1945) for decision-making under severe uncertainty through *worst-case analysis*. Robust optimization trades optimality for *conservatism* to guarantee feasibility of the design for the entire range of possible outcomes in future.

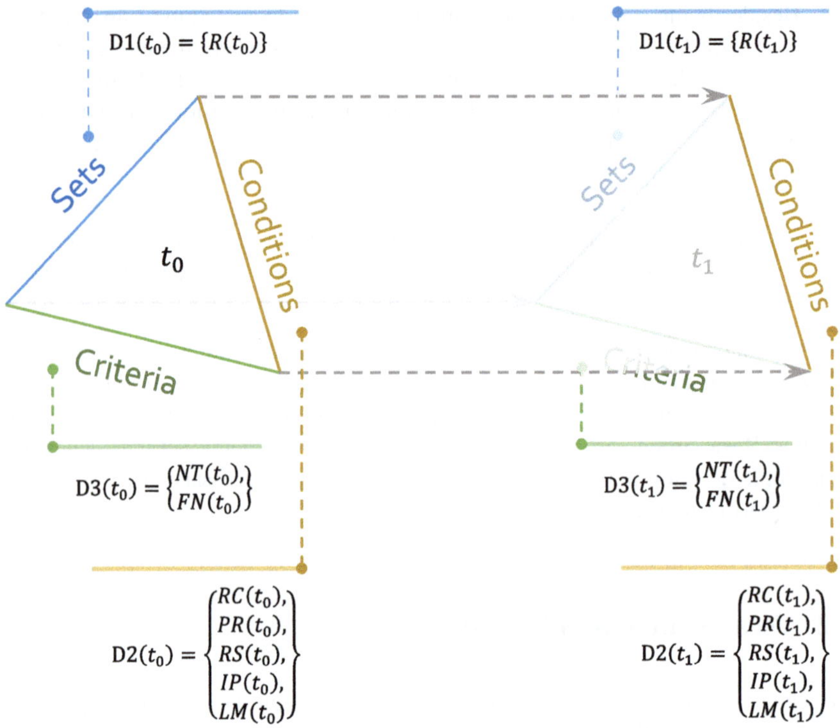

Fig. 6.1 Evolution of a best matching process in $\Delta = t_1 - t_0$ according to the PRISM taxonomy of best matching—"change" is defined as a process in Δ, where $\exists X \in \{1, 2, 3\}$, $\mathrm{DX}(t_0) \neq \mathrm{DX}(t_1)$

(2) *Fuzzy/stochastic optimization*. These approaches formulate uncertainty based on the current knowledge of the future state of the system. Formulation of uncertainty is performed in various ways; e.g., via probability (stochastic) or possibility (fuzzy) distributions. Popular approaches in this class include *chance-constrained*, *stochastic dynamic*, *flexible*, and *possibilistic programming*. Similar to robust optimization, these approaches do not contemplate the possibility of measurement, feedback, and modification of the design in real-time.

(3) *Real-time optimization*. When it is possible to measure, in real-time, the suitability of the model, parameters, and decision variables/objectives, and modify the design according to the provided measurements and feedback, RTO is usually the best alternative approach. Instead of relying on the current knowledge of the worst-case scenario or long-term state of the system, RTO continuously adapts and modifies the design models/decisions according to the feedback and real-time knowledge about the current/future performance of the system.

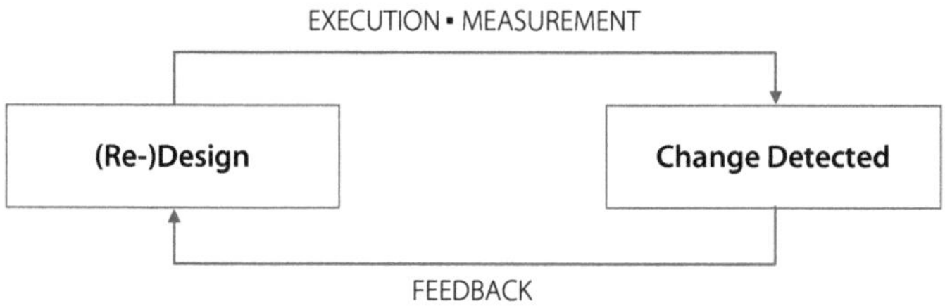

Fig. 6.2 RTO logic. Functions on the arrows require effective communication channels between distributed individuals and the matching designer(s)

The first two approaches are well established,[2] and thus are not covered in this book—also due to their disregarding of the possibility of real-time measurement, feedback, and modification, as valid functions in most practical best matching processes. The rest of this section is therefore dedicated to the basics of RTO methods for best matching in dynamic and uncertain environments (Fig. 6.2). RTO is inspired by the notion of *model predictive control* (MPC), an advanced process control method that has been extensively used in chemical and power industries for decades. The idea behind MPC is to optimize the system for the current period, while taking into account the anticipated state of the system in future periods (see, e.g., Chachuat et al. 2009). Based on the notion of MPC, two high-level RTO methods are presented next, with different ways of incorporating changes in the matching design.

6.2.1 Periodic Review Method

The periodic review method iteratively updates the design according to the detected changes during predefined periods of execution. That is, the system is designed at time t (based on the current knowledge of the system's characteristics and requirements), executed, and monitored for Δ time units, and then is redesigned (if needed) at time $t + \Delta$ with respect to the feedback information collected during Δ In a best matching process, if any changes are realized in the characteristics of sets, conditions, and/or criteria during Δ, the respective optimization algorithm must be run again to calculate the best match based on the new values at $t + \Delta$ (see Fig. 6.1). Below are the considerations and features of RTO with periodic review (Fig. 6.3):

- The runtime of the optimization algorithm (Γ) is a lower bound for the length of the review period Δ ($\Delta \geq \Gamma$).
- Longer/shorter review period Δ leads to *relatively* higher/lower communication and computational cost efficiency (Λ), and less/more optimal performance (Π) of the system.

[2]Interested readers are encourage to read Sahinidis (2004) and Ben-Tal et al. (2009).

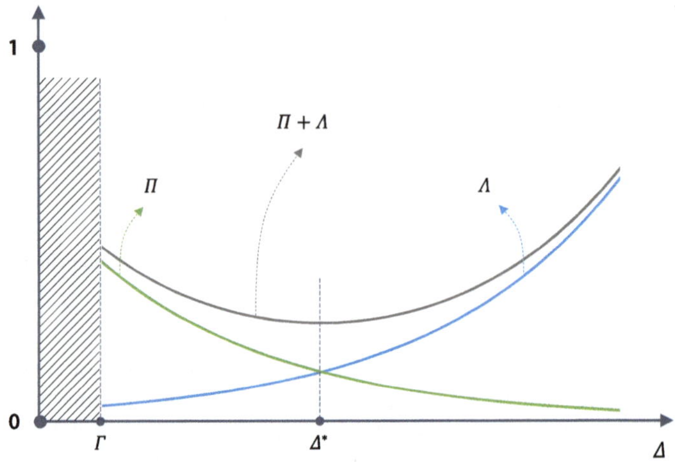

Fig. 6.3 Relationship between the length of the review period (Δ), communication and computational cost efficiency (Λ), and system's performance (Π). (Normalized $\Lambda, \Pi \in [0, 1]$; Γ: Runtime of algorithm; *: Optimality)

- With known relative importance of the communication and computational costs Λ and the system's performance Π, there is a unique optimal value for the length of the review period Δ.
- The length of the review period Δ can be fixed or variable.
- The communication times (for reporting and implementing changes) are assumed negligible.

Logic

A generic procedure for dynamic best matching with periodic review is presented below.

Step 1. *Initialization*. Identify the best optimization algorithm A for the given best matching problem. Estimate the runtime of the algorithm $\Gamma(A)$. Define the review period $\Delta \geq \Gamma(A)$. Set the current time to t.

Step 2. *Execution and measurement*. Execute A based on DX(t), $\forall X \in \{1, 2, 3\}$, implement the results, monitor and measure the system's performance during $[t, t + \Delta)$.

Step 3. *Feedback and update*. If DX(t) $\neq$ DX($t + \Delta$), update DX(t) $\leftarrow$ DX($t + \Delta$), $\forall X \in \{1, 2, 3\}$, according to the feedback collected during $[t, t + \Delta)$, set $t \leftarrow t + \Delta$, and return to Step 2. Otherwise, set $t \leftarrow t + \Delta$ and redo Step 3.

The above procedure can be applied to any best matching problem—and the respective optimization algorithm—that falls into this category. Any optimization algorithm similar to the algorithms discussed in Chap. 5 can be applied for periodic optimization of best matching problems (assuming its computation time is less than Δ). The key, however, is to effectively monitor, measure, and update the

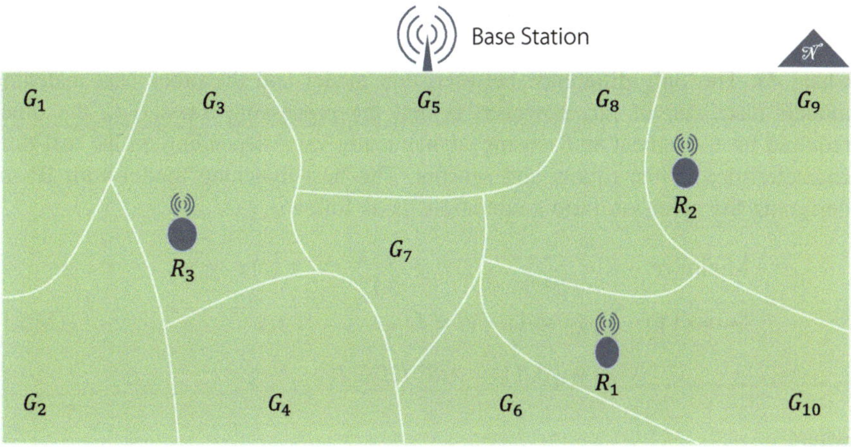

Fig. 6.4 Example of a precision farm with three ground robots (mobile) for real-time data collection from distributed sensors (stationary)

characteristics of the systems with respect to the three dimensions of best matching, and enable proper communication between the distributed individuals and the decision-maker(s). The following example highlights the notion of RTO with periodic review, and its application in best matching.

Example

Recall from Chap. 4 the example of precision agriculture with two types of robots for imaging and environmental monitoring (aerial robots), sensing, weeding, pruning, and applying fertilizers (ground robots) (see Fig. 4.5). Consider, as an example, a simple precision farm with three *unidentical* ground robots, which must continually move through the farm, collect data from 10 regions covered with several sensors, and transmit the data to the base station (Fig. 6.4). In each round, the robots leave the base station (their home position), and visit their respective regions through S-shaped movement towards south.[3] There are no limits on the number of regions allocated to each robot; this number, however, must be reduced, as the objective is to minimize the total movement and sensing times of the robots. Design of this precision agriculture system requires generalized two-sided best matching between the sets of regions $I = \{G_1, \ldots, G_{10}\}$ and robots $J = \{R_1, R_2, R_3\}$. The best matching problem is indeed *dynamic* and *uncertain*, due to high variability of the sensors' battery (energy) level and the robots' load, speed, battery (energy) level, location, assigned tasks, maintenance schedule, etc., which differ from one robot to another, and are also very difficult to measure in real-time.

[3]Note that without the assumption of S-shaped movement, the problem is a *multiple traveling salesman problem* (see Bektas 2006).

With the above definitions, the problem can be formulated as an *incapacitated* best matching model with interdependent preferences (IP) (see Model (M 3.9) in Chap. 3). The only difference between this model and the other best matching models discussed so far, however, is that the processing times (i.e., the times required for movement and sensing) dynamically vary, depending on the real-time characteristics of the robots and sensors. The best matching model with IP for designing the precision farm's operations is as follows:

$$\begin{aligned} \text{Minimize} \quad & Z = \sum_{i \in I} \sum_{j \in J} \tau_{ij}(t) \chi_{ij} + \sum_{i \in I} \sum_{j \in J} \sum_{i' \in I} \tau_{ij}(t) \alpha_{ii'} \chi_{ij} \chi_{i'j} \\ \text{Subject to} \quad & \sum_{j \in J} \chi_{ij} = 1, \quad \forall i \in I, \\ & \chi_{ij} \in \{0, 1\}, \quad \forall i \in I, j \in J. \end{aligned} \tag{M6.1}$$

where

$$\chi_{ij} = \begin{cases} 1, & \textit{if region i is processed by robot j} \\ 0, & \textit{if region i is processed by robot j} \end{cases} \quad \forall i \in I, j \in J,$$

$\tau_{ij}(t)$ denotes the estimated time to process region i by robot j based on the available knowledge of the system at time t, and $\alpha_{ii'}$ denotes the amount of increase (ratio) in the processing time of region i, if both regions i and i are assigned to the same robot. Parameters $\alpha_{ii'}$ are defined beforehand, based on the relative locations of the regions and the S-shaped movements of the robots towards south. Table 6.1 summarizes the current estimated processing times of the regions by each robot given that the robots directly visit the region after leaving the base station, along with the interdependent preferences of the regions (i.e., $\alpha_{ii'}$).

The RTO method with periodic review can be applied to this problem as follows:

Step 1. In Chap. 5, a capacitated version of best matching with IP was solved using ACO (Ant Colony Optimization). Hence, the same algorithm is

Table 6.1 Processing times estimates at $t = 0$ (min) and interdependencies between the regions' preferences (i.e., processing times)

i	$\tau_{ij}(0)$			$\alpha_{ii'}$									
	R_1	R_2	R_3	G_1	G_2	G_3	G_4	G_5	G_6	G_7	G_8	G_9	G_{10}
G_1	8	7	5	0	0	0	0	0	0	0	1.5	2.0	0
G_2	10	9	10	0	0	0	0.5	0	1.5	0.6	2.0	2.5	2.0
G_3	9	5	6	0	0	0	0	0	0	0	0.7	1.5	0
G_4	7	8	9	0.7	0	0.4	0	0	0.3	0	0.5	1.1	1.6
G_5	4	8	7	0	0	0	0	0	0	0	0	0	0
G_6	10	7	9	1.7	2.0	1.0	0.5	0	0	0	0.5	0.8	0.6
G_7	8	5	6	0.8	0	0.3	0	0	0	0	0.4	0.9	0
G_8	10	11	5	1.4	0	0.9	0	0	0	0.3	0	0	0
G_9	13	7	12	2.1	0	1.7	0	0	0	1.2	0	0	0
G_{10}	14	15	11	2.6	2.1	1.9	1.4	0	0.6	0.5	0	0.2	0

applied for solving the current problem.[4] The runtime of the ACO for this problem size is a few seconds (in the worst case). The length of the review period is set to $\Delta = 4$ min,[5] and the current time is set to $t = 0$ (see Table 6.1).

Step 2. Based on the current knowledge of the system at $t = 0$ (i.e., Table 6.1), the ACO is run, and yields the following matching results:

	G_1	G_2	G_3	G_4	G_5	G_6	G_7	G_8	G_9	G_{10}
R_1	3	4	2	–	1	–	–	–	–	–
R_2	–	–	–	2	–	3	1	–	–	–
R_3	–	–	–	–	–	–	–	1	2	3
Z^*	87.1 min									

The results above indicate the regions assigned to each robot along with their sequence of operations (e.g., R_1 must process G_5, G_3, G_1, and G_2, in that order). The system's performance must then be monitored and measured with this design during period $[0, 4)$.

Step 3. Now suppose the processing time estimates are updated at time $t = 4$—based on the feedback collected during $[0, 4)$—as follows:

Table 6.2 Processing times estimates at $t = 4$ (min)

	$\tau_{ij}(4)$		
	R_1	R_2	R_3
G_1	12	5	5
G_2	13	6	10
G_3	12	3	6
G_4	11	6	9
G_5	6	7	7
G_6	12	4	9
G_7	13	5	6
G_8	14	9	5
G_9	16	6	12
G_{10}	15	12	11

Compared to Table 6.1, the updated processing time estimates in Table 6.2 indicate that the actual performances of robots R_1, R_2, and R_3 (based on the feedback collected during $[0, 4)$) are respectively lower, higher, and the

[4]Note that best matching with IP is a quadratic assignment problem, which can be solved by various exact (e.g., branch-and-bound), heuristics/metaheuristics (e.g., genetic algorithm, tabu search). Details of the ACO have not been presented for the sake of brevity. (Review Chap. 5 for more detailed explanation of ACO.).

[5]Note that the optimal length of Δ can be defined based on the tradeoff shown in Fig. 6.3.

same as their expected values at $t = 0$. (The shaded cells in Table 6.2 denote the regions that are in process by the robots at $t = 0$.) With these updates, it is likely that the original matching decisions are not optimal anymore. According to Table 6.2, at $t = 4$, R_1, R_2, and R_3 are still processing G_5, G_7, and G_8, respectively. Hence, the remaining regions must be re-matched to the robots with respect to the updated processing time estimates.

Step 2. (*Revisited*) Based on the updated knowledge of the system at $t = 4$ (i.e., Table 6.1), the ACO is run, and yields the following matching results[6]:

	G_1	G_2	G_3	G_4	G_5	G_6	G_7	G_8	G_9	G_{10}
R_1	2	3	–	–	1	–	–	–	–	–
R_1	–	–	2	3	–	4	1	–	–	–
R_1	–	–	–	–	–	–	–	1	2	3
Z^*	90.9 min									

The updated matching results indicate a small change in the new design compared to the previous one: G_3 must be processed by R_2 instead of R_1.

The same procedure is repeated at $t = 8, 12, 16, \ldots$ according to the updates in the processing time estimates, rooted in the evolution of the system's characteristics and requirements in real-time. Periodic review is the most basic RTO method. It is essentially based on recurrent execution of optimization algorithms. This method is effective for best matching applications with relatively more stable domains (i.e., low frequency of changes), and cases where the runtime of the optimization algorithm is negligible. For more dynamic and complex applications, however, more advanced RTO methods must be developed and deployed, as described next.

6.2.2 *Continuous Review Method*

The periodic review method is applicable in cases where $\Gamma < \Delta$, i.e., when the runtime of the algorithm is lower than the length of the review period. The significant *complexity* and *dynamicity* of many practical best matching instances, however, reverse this inequality (i.e., $\Gamma \gg \Delta$) by significantly increasing Γ and decreasing (the best value of) Δ, respectively, thus reducing the suitability of the periodic review method. The logic of the continuous review method—instead of accumulating feedback during Δ and applying changes at the end of each period—is

[6]The shaded cells refer to the processed/in-process regions—the matching decisions associated with these regions cannot be changed.

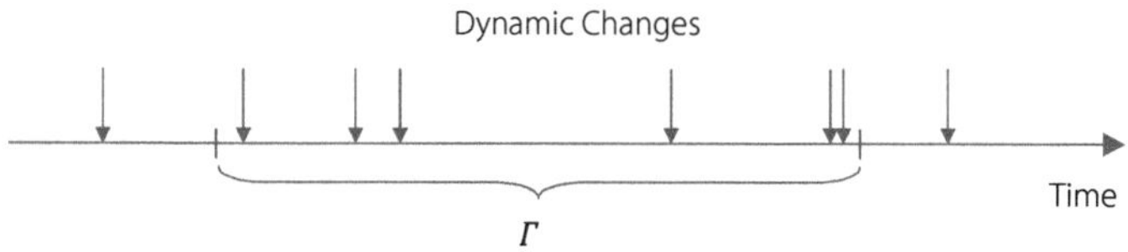

Fig. 6.5 Dynamic occurrence of changes (vertical arrows) during one complete runtime of optimization algorithm—the continuous review method modifies and reruns the algorithm with respect to each single change, while having the best solution ready at any point in time

to constantly detect, analyze, and (if and when warranted) adapt to each single change that takes place after the initial design[7] (Fig. 6.5). The goal of the continuous review method is to *always have the best possible solution ready*, while constantly looking for better solutions in the background based on real-time changes (and trending changes) in the system. RTO with continuous review facilitates *responsive adaptation* to dynamic changes based on the following basic assumptions:

- The communication times (for reporting and implementing changes) are negligible.
- Only one significant change (i.e., warranting adaptation) can take place at a time.

Adaptation to real-time changes requires certain procedures for rapid transition from the old (before the change) to the new (after the change) problem. These procedures are typically heuristic, due to the diverse and domain-specific nature of changes in different best matching instances. The general logic of *adaptation heuristics* for RTO with continuous review is described next.

Logic

A substantial number of optimization algorithms, in general, and best matching algorithms, in particular, comprise an *initialization* element (e.g., genetic algorithm, ant colony optimization), tabu search, branch-and-bound method). The role of the initialization element is to generate an initial set of solutions for a new problem instance, which are then improved through certain search mechanisms towards the optimal solution. Naturally, the initialization element significantly influences the computational efficiency of optimization algorithms—the higher the quality of the initial solution(s), the higher the chance of finding better solutions faster. The implication of this property for RTO lies in the following logic: *After each change, the old problem turns into a new problem; however, since only one significant change can take place at a time, there is a high chance of creating a high quality initial solution to the new problem, based on the best existing solution to the old*

[7]See Moghaddam and Nof (2015a, b, 2016) for detailed modeling and analysis of the RTO with continuous review, with applications in supply network design, enterprise collaboration, and assembly line design.

problem. Research on RTO, in general, and on adaptation heuristics, in particular, is in its infancy stages, and generally relates to process optimization applications in chemical and power industries (see, e.g., Pistikopoulos et al. 2002; Zhang et al. 2002; Chachuat et al. 2009). Thus, there is not much literature to discuss, regarding dynamic best matching processes with respect to the three dimensions of the PRISM taxonomy of best matching (see Fig. 6.1). In the following, a number of adaptation heuristics are presented, concerning the changes in the number of individuals in the system (Moghaddam and Nof 2015a); i.e., *association* of a new individual to the system or *dissociation* of an existing individual from the system. This dynamic behavior is linked to the *association/dissociation principle* of the collaborative control theory.[8]

Association/Dissociation of Individuals.

The principle of association/dissociation (*a.k.a.*, the join/leave/remain principle) addresses the dynamic changes in distributed systems such as enterprise alliances, sensor clusters, robot teams, modular systems, etc. in terms of system size and configuration (Chituc and Nof 2007; Yoon and Nof 2011). Examples include

- Arrival/departure of jobs to/from a production cell, or a supply network;
- Purchasing or salvaging production equipment or transportation facilities;
- Introduction/declination of a product in a market;
- Emergence/saturation of a market;
- Activation/deactivation of sensors in a sensor array;
- Allocation of interns/students to hospitals/schools;
- Establishing a new school/hospital;
- Temporary or permanent loss of a team member or leader;
- Opening or closing a line in a service system such as a bank, or energy distribution grid.

A graphical representation of this type of behavior is shown in Fig. 2.8, Chap. 2, for generalized two-sided best matching $M: 1//+, OS$ between sets I and J (see Model (M 3.2) in Chap. 3). The same example is used here for explanation of the adaptation heuristics associated with RTO with continuous review.[9] For this specific case, there are four possible scenarios related to (1) association or (2) dissociation of the elements of (3) set I or (4) set J. The adaptation heuristics for these

[8]For detailed information about the collaborative control theory and its design principles, interested readers are encouraged to read "Revolutionizing Collaboration through e-Work, e-Business, & e-Service" by Nof et al. (2015).

[9]The presented heuristics are based on a recent work by Moghaddam and Nof (2015a) on generalized best matching, and can be extended to other instances of best matching in a similar manner.

scenarios are presented below, where the best current solution (i.e., before change) is denoted by β.[10]

(1) *Association of the elements of I* (Fig. 2.8a). After association of a new individual u to set I (i.e., $I \leftarrow I \cup \{u\}$), append a new cell corresponding to u to the end of β, whose entry represents the match of k, $v \in J$. In the modified solution $\hat{\beta} = [\beta\, v]$, v must satisfy the following condition:

$$P_{uv} = \max_{j \in L}\{P_{uj}\}, \tag{6.1}$$

where $L = \{j : M_j > \sum_{i \in I} \chi_{ij}\}$ (see Model (M 3.2) for details). If $L = \emptyset$, set $v = 0$, which implies that u remains unmatched (i.e., $\chi_{uj} = 0, \forall j \in J$). The idea behind this heuristic is to match the new individual $u \in I$. to its best possible match in set J that has extra capacity.

(2) *Dissociation of the elements of I* (Fig. 2.8b). After dissociation of an existing individual u. from set I (i.e., $I \leftarrow I \backslash \{u\}$), remove the cell corresponding to u from β to generate the modified solution, i.e., $\hat{\beta} = \left[\beta_1 \cdots \beta_{u-1}\, \beta_{u+1} \cdots \beta_{|I|}\right]$. The match of u in set J will end up with an extra capacity after dissociation of u, which is not expected to result in significant decrease in the quality of the modified solution. Hence, this heuristic simply leaves the modified solution as is, without making any changes in the matching of the remaining individuals.

(3) *Association of the elements of J* (Fig. 2.8c). After association of a new individual v to set J (i.e., $J \leftarrow J \cup \{v\}$), develop the modified solution $\hat{\beta}$ as follows:

   a. Sort the elements of set I in ascending order of P_{iv} and place them in a temporary set K.
   b. Pick the first element of and denote it by u.
   c. If $P_{iv} > P_{ij}$, where $j = \hat{\beta}(u)$ (i.e., j is the current match of u), set $\hat{\beta}(u) = v$ (i.e., match. to v), and update $K \leftarrow K \backslash \{u\}$ and $M_v \leftarrow M_v - 1$. If $T = \emptyset$ or $M_v = 0$ stop and report the modified solution $\hat{\beta}$. Otherwise, return to Step (b).

   The idea behind this heuristic to identify the *least satisfied* elements of set I, and re-match them, one at a time, to the new element $v \in J$, if this change increases the overall quality of the solution. The procedure stops when there are no such elements in I, or when the new individual v runs out of capacity.

(4) *Dissociation of the elements of J* (Fig. 2.8d). After dissociation of an existing individual v from set J (i.e., $J \leftarrow J \backslash \{v\}$), develop the modified solution $\hat{\beta}$ as follows:

---

[10] In all the four adaptation heuristics, it is assumed that β (and each solution, in general) is encoded as an array of length $|I|$, where the entries represent the elements of set J matched to each $i \in I$ (see Fig. 5.5 in Chap. 5).

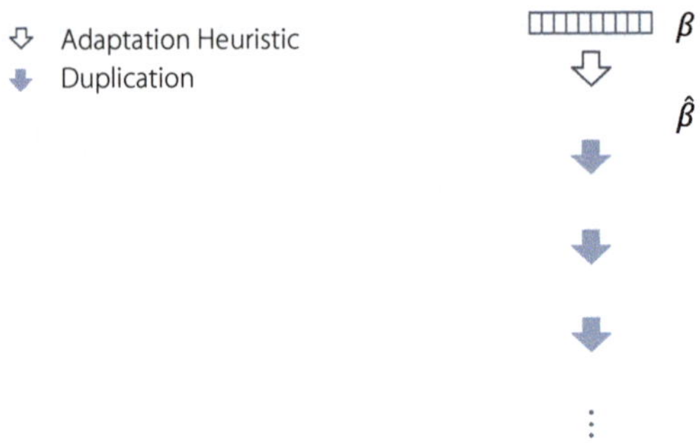

Fig. 6.6 Generation and "Mitosis-like" duplication of the modified solution (chromosome) via reproduction functions of genetic algorithm (e.g., mutation, crossover) for generating the first population of solutions (after Moghaddam and Nof 2015a)

a. Set $\hat{\beta} = \beta$.
b. Define $K = \{i : \chi_{iv} = 1\}$ and $L = \left\{j : M_j > \sum_{i\in I} \chi_{ij}, j \in J\backslash\{v\}\right\}$. If $K = \emptyset$ or $L = \emptyset$, go to Step (d). Otherwise, proceed to the next step.
c. Find $i \in K$ and $j \in L$, where $P_{ij} = \max_{k\in K, l\in L}\{P_{kl}\}$. Set $\hat{\beta}(i) = j$, and return to Step (b).
d. IF $K = \emptyset$ set $\hat{\beta}(i) = 0, \forall i \in K$.

The above procedure matches the corresponding elements of v in set I to other elements of set $J\backslash\{v\}$ that have extra capacities (if any), in a way that the overall satisfaction is maximized. In some cases, however, the available capacity of set $J\backslash\{v\}$ is not sufficient, and thus some elements of I corresponding to v may remain unmatched due to capacity shortage.

The heuristics described above do not guarantee optimality of the modified solution $\hat{\beta}$, but achieving satisfactory quality in reasonable amount of time. The modified solution can then be used as a good starting point to initiate the search towards the optimal solution or generate a high-quality initial population (for population-based methods) through various heuristic operators (e.g., Fig. 6.6). These (and similar) heuristics significantly improve the *speed* and *quality* of adaptation to changes caused by association/dissociation of individuals to/from the system (Moghaddam and Nof 2015a).

Example

Sensor networks and their applications are growing, substantially, to address the emerging needs for autonomous distributed monitoring and control of processes (e.g., characteristics such as position, velocity, humidity) with better fault tolerance, accuracy, timeliness, cost effectiveness, and scalability, compared to traditional single-sensor systems.[11] For energy and communication efficiency, sensor networks are usually designed with *multi-layer* architectures, where individual sensors form

[11]Interested readers are encouraged read "Sensors and Sensor Networks" by Jeong (2009) in the *Springer Handbook of Automation* (Nof, *Ed.*), and Jeong and Nof (2009).

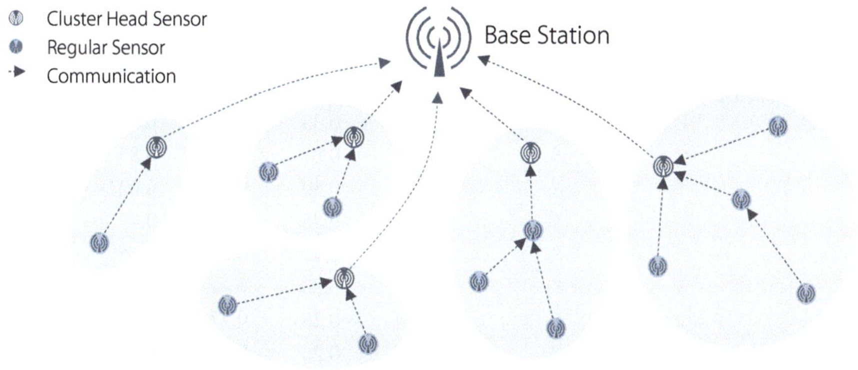

Fig. 6.7 Example of a sensor network with single- and multi-hop clusters—sensors communicate with the base station by transmitting information to the cluster head (single-hop) or another sensor in the cluster (multi-hop)

clusters and communicate information with the higher layer (e.g., base station) via a *cluster head* (Fig. 6.7). Due to the relatively short lifespans of sensors in terms of energy level (especially micro-/nano-sensors), configuration of sensors networks is highly dynamic, involving association of new sensors to the network and temporary/permanent dissociation of the existing ones, including the cluster heads. Hence, design of sensor networks with such properties can be recast as a generalized two-sided matching problem with *emergent sets*, i.e., M: $1//+, OS/ES$ (see Chap. 2), where sets I and J respectively refer to the set of regular sensors and the set of cluster heads (see Fig. 6.7). Table 6.3 summarizes the input data for an example where 20 sensors must form 8 disjoint and capacitated clusters with predefined and fixed cluster heads, i.e., $|I| = 12$ and $|J| = 5$.

The initial problem with the information provided at time $t = 0$ is a two-sided matching problem (see Model (M 3.2) in Chap. 3), which can be solved using most of the algorithms described in Chap. 5. Due to the relatively small size of the problem, however, Model (M 3.2) is solved using the mixed-integer linear programming toolbox of MATLAB® (`intlinprog`), which yields the following optimal solution with $Z^* = 8.12$.

	S_1	S_2	S_3	S_4	S_5	S_6	S_7	S_8	$S_9.$	S_{10}	S_{11}	S_{12}
β	C_5	C_4	C_3	C_1	C_2	C_2	C_2	C_3	C_5	C_5	$C_2.$	C_4

It implies that sensor S_1 is matched to cluster C_5, sensor S_2 is matched to cluster C_4, and so on. With β as the current solution, the following real-time changes are addressed by the aforementioned adaptation heuristics, as shown in Fig. 6.8.

Table 6.3 Mutual preferences of sensors and cluster heads, and cluster heads' capacities[a]

	Cluster heads				
Sensors	C_1	C_2	C_3	C_4	C_5
S_1	0.60	0.06	0.65	0.53	0.77
S_2	0.28	0.35	0.43	0.42	0.56
S_3	0.47	0.87	0.88	0.83	0.59
S_4	0.56	0.50	0.16	0.02	0.48
S_5	0.39	0.75	0.36	0.42	0.03
S_6	0.18	0.72	0.37	0.45	0.18
S_7	0.67	0.98	0.41	0.56	0.32
S_8	0.44	0.28	0.77	0.65	0.73
S_9	0.28	0.26	0.34	0.34	0.55
S_{10}	0.34	0.24	0.43	0.45	0.67
S_{11}	0.25	0.42	0.20	0.24	0.40
S_{12}	0.29	0.72	0.52	0.63	0.18
Capacity	2	4	2	3	3

[a]Based on sensors' proximity, energy level, and communication protocols. Capacity represents the maximum number of sensors that a cluster head can connect to the base station

(1) Sensor S_{13} joins the network with the following mutual preferences with the cluster heads:

$j.$	$C_1.$	$C_2.$	C_3	$C_4.$	$C_5.$
$P_{S_{13},j}$	0.43	0.50	0.70	0.40	0.14

The best cluster head for S_{13} (in terms of preference score) must be selected out of the ones with extra capacity, i.e., $L = \{C_1, C_4\}$.

(2) Sensor S_{10} leaves the network.

(3) Cluster head C_6 joins the network with the capacity of 2 and the following mutual preferences with the sensors:

i	$S_1.$	$S_2.$	S_3	$S_4.$	S_5	$S_6.$	$S_7.$	$S_8.$	$S_9.$	$S_{10}.$	S_{11}	$S_{12}.$
$P_{i,C_6}.$	0.55	0.34	0.45	0.37	0.77	0.63	0.20	0.54	0.48	0.49	0.08	0.52

The sensors that prefer $C_6.$ to their current cluster head (if any) leave their current cluster and join the cluster of C_6, considering its capacity availability.

(4) Cluster head C_5 leaves the network. Its corresponding sensors, i.e., $K = \{S_1, S_9, S_{10}\}$, must then be matched to the remaining cluster heads with available capacity, i.e., $L = \{C_1, C_4\}$, considering their mutual preferences with the temporarily unmatched sensors.

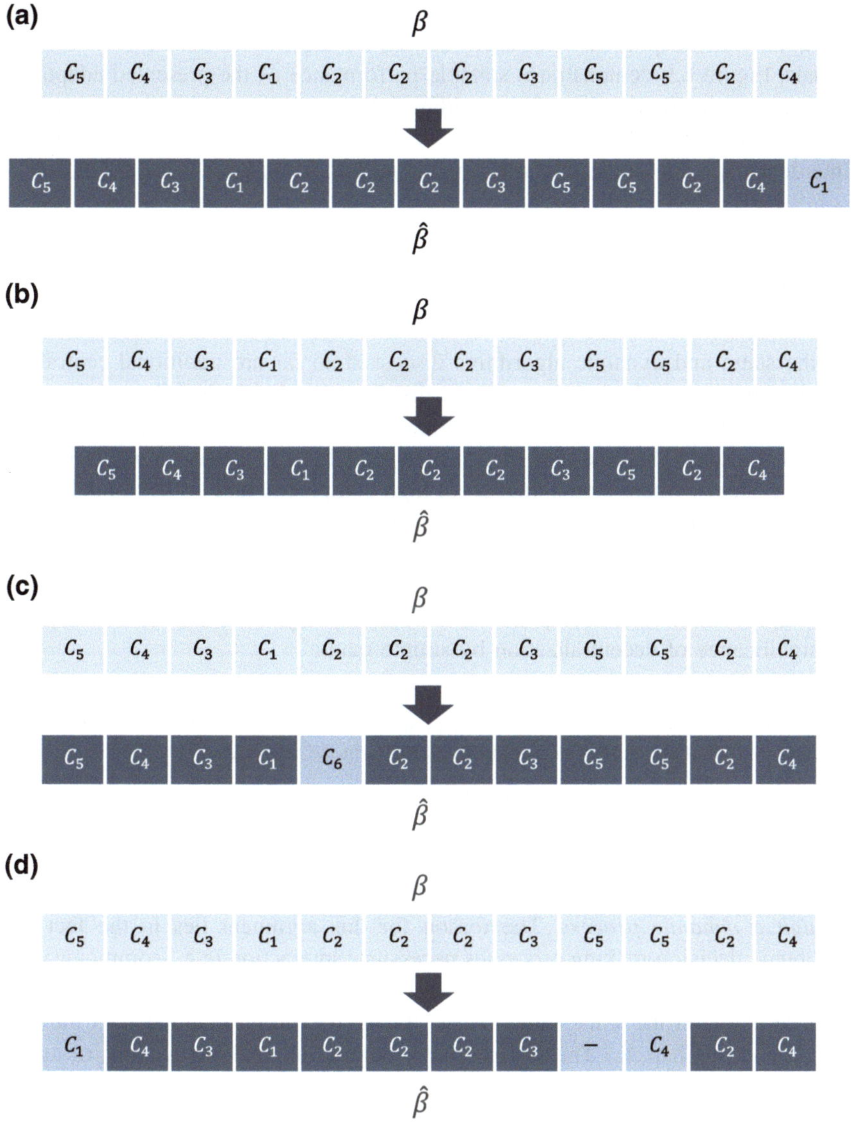

Fig. 6.8 Generation of modified solutions (β) using the adaptation heuristics for association/dissociation. **a** Association of S_{13}. **b** Dissociation of S_{10}. **c** Association of C_6. **d** Dissociation of C_5 (S_9 remains unmatched due to the overall lack of capacity)

Application of the adaptation heuristics in the above illustrative example may seem trivial, because the problem has a relatively standard structure (classic two-sided generalized matching), a its optimal solution can be found in a fraction of a second due to its small size. In more complex and unstructured instances, however, the presented (and similar) heuristics show significant impact on the speed and quality of adaptation to changes caused by association/dissociation of individuals. Figure 6.9 demonstrates suitable performance of the presented adaptation heuristics, coupled with genetic algorithm for handling real-time changes in four different instances of generalized resource-constrained best matching with IP (Interdependent Preferences) and emergent se; i.e., $M{:}\,1/RC, IP/+, OS/ES$.

6.3 Distributed Control

All the static and dynamic algorithms discussed so far are traditional centralized decision-making methods, where all the distributed individuals follow a single set of decisions made by a central decision-maker. As discussed in Chap. 4, however, this is not always the case (see Fig. 4.2); especially in modern manufacturing and service systems, where a combination of high complexity and artificial intelligence is increasingly making a paradigm shift towards *decentralization* (see Fig. 4.3). The implication, in simple terms, is that instead of one central decision-maker, a set of distributed decision-makers interact with each other and design the system together —through collaboration, competition, or both. Van Dyke Parunak (1997) explains the significance of decentralization by simple math:

> One hundred agents, each with ten behaviors, require the programming of only 1000 individual behaviors; yet provide a behavior space on the order of 10^{1000}, a number far larger than the total number of elementary particles in the universe.

Agents are the autonomous, distributed decision-makers. The systems of rules that "orchestrate" agents, enable them to interact, be productive, and achieve their goals are referred to as *protocols*. A remarkable property of such settings, where multiple decision-makers (agents) are involved, is that *matching is essentially done through a dynamic process*. The reason for this argument lies in the fact that distributed decision-making processes necessitate interaction (e.g., communication, negotiation, information exchange) between agents, and not all of them reach their local decisions at the same time. (These issues are further elaborated in the next section and in Chap. 8.) The notions, basic definitions, and applications of agents and protocols for distributed best matching are discussed next.

6.3.1 Multi-agent Systems

Agents are computational elements with certain levels of *intelligence* and *autonomy* capable of responding to expected and unexpected events in their "local" environment. Multi-agent systems enable the *localization* of design and decision-making via loosely coupled networks of agents that are:

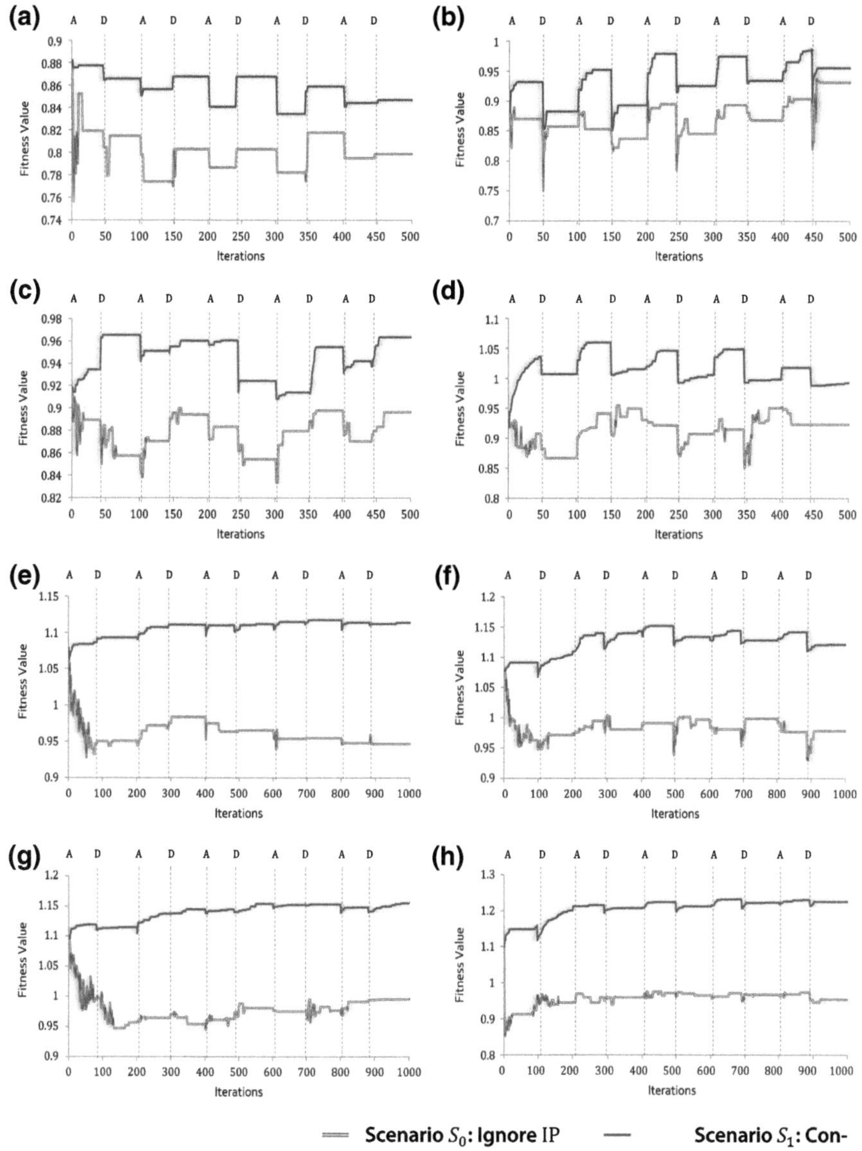

Fig. 6.9 Evolution of capacitated clusters through association (A)/dissociation (D) of I (**a**, **c**, **e**, **g**) and J (**b**, **d**, **f**, **h**) in four instances of $M: 1/RC, IP/+, OS/ES$; (**a**, **b**): $SZ(20, 8)$; (**c**, **d**): $SZ(40, 18)$; (**e**, **f**): $SZ(60, 25)$; (**g**, **h**): $SZ(80, 30)$, where $SZ(|I|, |J|)$. *Observations*: high impact of IP on the overall satisfaction and the possibility of "erroneous evolution", if disregarded (S_0); rapid adaptation to changes via the adaptation heuristics and genetic algorithm; lower overall impact of single changes as the system size grows (after Moghaddam and Nof 2015a)

- *Social.* Interact with other agents and with their environment through certain communication languages and protocols.
- *Redundant.* Enhance *dependability*, i.e., reliability and fault tolerance, of the systems by operating as *teams* rather than as individuals.[12]
- *Independent.* Take action within their local environment without any need for higher-level control.
- *Proactive.* Dynamically predict and respond to the signals received from other agents and their environment.

The aforementioned traits certainly necessitate agents to be intelligent and autonomous, in order to augment their social behaviors, and achieve individual and common goals. Methods and applications of agent-based systems are expanding intensively, making them omnipresent in almost every manufacturing and service industry; e.g., transportation and supply network design, production planning and scheduling, healthcare operations, air traffic control, emergency response planning, e-Commerce, and e-Entertainment (Jennings et al. 1998; Nof 2000).

Interaction of agents—with each other and with their environments—is the key element for survival, evolution, and goal accomplishment of agents. It lies in their sociability and redundancy, supported by their intelligence (for high quality interaction) and autonomy (for timely interaction). As discussed in Chap. 4, the interaction between two agents can be defined in two different manners:

(1) *Hierarchy* versus *Heterarchy.* Is the relationship between the agents *vertical* or *horizontal*? If the agents are not at the same organizational (authority) level and have master-slave relationship, their interaction is defined as *hierarchical.* If the agents are at the same level with no definite master-slave relationship, their interaction is defined as *heterarchical* (see Fig. 4.3).
(2) *Collaboration* versus *Competition.* Are the agents seeking *common* or *conflicting* goals? If the agents are trying to accomplish a set of common or complementary goals, their interaction is defined as *collaboration.* If the goal accomplishment of one agent is likely to make long-term negative impact on the other agent's goals, their interaction is defined as *competition* (see Fig. 4.6).

The next step after defining the agents, their responsibilities, and nature of interactions, is to design protocols that enable efficient orchestration and administration of distributed agents and their corresponding tasks. Before discussing the details and mechanics of such interaction protocols, however, an agent-based shop floor control is presented below as an illustrative example for better understanding of the notion of multi-agent systems.

[12]This trait is related to the CCT (Collaborative Control Theory) principle of *collaborative fault tolerance*, which states that *a team of weak agents always outperforms a strong individual agent in long-term* (see Liu and Nof 2004; Jeong 2006; Nof 2007).

Fig. 6.10 An agent-based shop floor with three types of agents for parts, machines, and AGVs (☺: Agent)

Example

Consider an agent-based manufacturing shop with three types of agents for parts, machines, and material-handling equipment (Fig. 6.10). The shop operates with three machines M_1, M_2, and M_3, and two AGVS (Automated Guided Vehicles) G_1 and G_2. The problem is to find the best three-sided match between the sets of parts, machines, and AGVs over time and in a distributed environment. The following scenario is an example that shows how such best matching processes can be designed using the notion of multi-agent systems:

(1) Part P_1 arrives in the system and part agent Ω^{P_1}. is created accordingly.
(2) Ω^{P_1}. submits its information (e.g., type, quantity, due date) to machine agents Ω^{M_1}, Ω^{M_2}, and Ω^{M_3}, and requests their information (e.g., availability, queue length, processing time).
(3) Ω^{M_1} reports "M_1 out of order". Ω^{M_2}. and Ω^{M_3} report availability, but with significantly different processing information, since M_3 is "overloaded".
(4) Ω^{P_1} selects M_2 as the best match for P_1, and requests AGV agents Ω^{G_1} and Ω^{G_2} their information (e.g., current location, anticipated waiting time, eligibility).
(5) Ω^{G_1} and Ω^{G_2} both prepare and submit proposals to Ω^{P_1}.
(6) Ω^{P_1} selects G_1 as it is a better match for P_1.

This is an example of a multi-agent system with purely heterarchical relationships between the agents—they all are at the same organizational level and there are no master-slave relationships between them. From the type of negotiations, it can be argued that the machines have collaborative relationships as they all prioritize the overall benefit of the shop relative to their individual benefits—by trying to balance the workload, and reduce the waiting time and congestion, rather than merely (greedily) offering to process the part. The AGVs, on the other hand, are competitors as they both offer to carry P_1, even though G_2 is currently busy and P_1 has to wait if matched to it. This is indeed a highly simplified scenario of real-life situations, where large numbers of agents must constantly communicate and negotiate to accomplish their local and system-wide goals. Design of such systems is essentially more complicated than the few steps taken in this example, and thus, must be carried out in more systematic ways via carefully designed interaction protocols.

6.3.2 Interaction Protocols

Interaction protocols provide efficient platforms for agents to communicate, work in harmony with each other and with their environments, and accomplish their individual and common goals. These protocols are also referred to as *task administration protocols* (TAP) (Nof 2003; Ko and Nof 2012); because regardless of the context and application, every distributed system can be abstracted as a network of *resources* (e.g., agents, machines) deployed for processing a set of *tasks*. In the aforementioned agent-based shop floor system, for example, agents are responsible for tasks such as information exchange (e.g., task data, machine status), negotiation (e.g., proposal submission by the AGVs), and decision-making (e.g., selecting the best machine/AGV). Accomplishment of each single task, however, is directly influenced by the quality and synchrony of interactions among agents, which can be augmented via interaction protocols. This section introduces TAP as a generic interaction protocol for the design and administration of *collaborative* multi-agent systems, particularly, for best matching purposes.

Logic

To comprehend the notion of distributed control, in general, and TAP, in particular, as well as the transition from the optimization models discussed so far to interaction protocols in distributed settings, consider the classic generalized resource constrained matching problem $M{:}\,1/RC/+,OS$ between a set of tasks I and a set of machines J.[13] In a centralized setting, this problem is formulated as follows (see Model (M 3.5) in Chap. 3):

[13]This problem is also known in literature as the *generalized assignment problem* (see Pentico 2007).

$$
\begin{aligned}
\text{Maximize} \quad & Z = \sum_{i \in I} \sum_{j \in J} P_{ij} \chi_{ij} \\
\text{Subject to} \quad & \sum_{i \in I} \theta_{ij} \chi_{ij} \le T_j, && \forall j \in J, \\
& \sum_{j \in J} \chi_{ij} = 1, && \forall i \in I, \\
& \chi_{ij} \in \{0, 1\}, && \forall i \in I, j \in J.
\end{aligned} \tag{M6.2}
$$

The objective of Model (M 6.2) is to maximize the overall satisfaction of preferences P_{ij} (e.g., profit; task-machine compatibility), considering (1) the total available times of the machines, i.e., T_j, where θ_{ij} denotes the processing time of task i on machine j, and (2) the fact that each task must be processed by exactly one machine. This is a standard matching problem and can be solved using most of the optimization algorithms discussed in Chap. 5 (e.g., branch-and-bound; genetic algorithm; tabu search). Even if there are additional conditions, such as precedence relationships between the tasks (see Model (M 3.5) in Chap. 3) or resource sharing between the machines (see Model (M 3.6) in Chap. 3), the problem can still be solved by relatively minor modifications in the above-mentioned (or similar) algorithms. In a distributed setting, however, the design process is relatively more challenging, involving three sub-protocols of TAP for *task requirement analysis*, *shared resource allocation*, and *synchronization*, (following Ko and Nof 2012) as described next.[14]

Task Requirement Analysis Protocol (TRAP)

TRAP is activated when task i arrives in the system at time t Task agent Ω^i applies TRAP to define the precedence relationships and interdependencies of i with other tasks in the system, and evaluate its *relative priority*. If the priority of task i is less than the priority of all other tasks in the system, it is simply added to the end of the task queue following the FIFO (First-In First-Out) discipline. Otherwise, the sequence of the tasks in the queue is updated according to their new priority values. This update may also be applied to the task(s) already in process by preemption, if i is a high-priority urgent task that must be processed right away. Task agent Ω^i defines task i at time t as

$$
i(t) = \langle \tau_i(t), \theta_{ij}(t), \delta_i(t), \pi_i(t), P_{ij}(t) \rangle, \tag{6.2}
$$

where $\tau_i(t)$, $\theta_{ij}(t)$, $\delta_i(t)$, $\pi_i(t)$, and $P_{ij}(t)$ denote its type, processing time on machine $j \in J$, due date, profit, and preference (e.g., relative priority) for machine $j \in J$, respectively.[15]

[14]Note that this is just an example, and interaction protocols can be designed in many different ways. Interested readers are encouraged to read the seminal article on *Contract Net Protocols* by Smith (1980), which introduces the basics of collaborative task sharing in multi-agent systems.

[15]This time-based definition implies that all characteristics of task i are (or can be) subject to change over time, and thus are continuously monitored and updated by task agent Ω^i.

Table 6.4 Sensitivity of the preference of i to other main characteristics

	Preference of i	Policy
Profit of i ↑	↑	Higher economic throughput
Processing time of i ↑	↓	Shortest processing time first
Time until due date of i ↑	↓	Earliest due date first
Remaining processing time of k ↑	↑	Higher fulfillment rate

TRAP calculates the real-time preference of task i for machine $j \in J$—relative to the other unassigned tasks and as a function of their dynamic characteristics—as follows:

$$p_{i \to j}(t) = w_1 \cdot \frac{\pi_i(t)/\theta_{ij}(t)}{\sum_{l \in U}\left(\pi_l(t)/\theta_{lj}(t)\right)} + w_2 \cdot \left(1 - \frac{\delta_i(t) - t}{\sum_{l \in U}(\delta_l(t) - t)}\right) + w_3 \cdot \frac{\breve{\theta}_{kj}(t)}{\theta_{kj}(t)}, \quad \forall i \in U, j \in J, \tag{6.3}$$

where w_n is the relative weight of each term ($\sum_{n=1}^{3} w_n = 1$); U is the set of unassigned tasks (including i); k is the task already in process by machine j; and $\breve{\theta}_{kj}(t)$ is the *remaining* processing time of task k. According to Eq. (6.3), the dynamic characteristics of task i influence its preference score as summarized in Table 6.4. Task agent Ω^i dynamically calculates and updates $p_{i \to j}(t)$, $\forall j \in J$, and shares this information with its *most preferred* machine agent(s) in the system, leading to the activation of the second sub-protocol of TAP.

Shared Resource Allocation Protocol (SRAP)

SRAP is activated when machine agent Ω^j receives $p_{i \to j}(t)$ from task agent Ω^i. Each machine agent Ω^j then evaluates the proposals (if any), and prepares a *bid* (i.e., calculates its preference score) for each task i at time t as follows:

$$p_{j \to i}(t) = w_1 \varphi_{ik} + w_2(1 - \xi_k) + w_3 \cdot \left(1 - \frac{\sum_{l \in Q} \theta_{lj}(t)}{T_j}\right), \quad \forall i \in U, j \in J, \tag{6.4}$$

where w_n is the relative weight of each term ($\sum_{n=1}^{3} w_n = 1$); Q is the set of tasks in the queue of machine j; k is the task already in process by machine j; $\varphi_{ik} = 1$, if $\tau_i(t) = \tau_k(t)$, and $\varphi_{ik} = 0$, otherwise; ξ_k is the penalty of preempting task k ($0 \le \xi_k \le 1$); and T_j is the total available time of machine j. According to Eq. (6.4), the bid prepared by machine j is a function of three factors shown in Table 6.5. Machine agent Ω^j prepares and submits the bid to task agent Ω^i for evaluation. Task agent Ω^i then evaluates the situation based on its initial preference scores (i.e.,

Table 6.5 Sensitivity of the bi of j (preferences) to other main characteristics

	Preference of j	Policy
Tasks i and k are the same type	$\uparrow$	Save setup time and frequency
Penalty of preempting $k\uparrow$	$\downarrow$	Reduce rework
Total workload of $j\uparrow$	$\downarrow$	Reduce wait time in the system

$p_{i\to j}(t)$, $\forall j \in J$) and the received bids ($p_{j\to i}(t)$, $\forall j \in J$), and calculates the overall preference value for each j as follows[16] (see Eq. (2.1) in Chap. 2):

$$P_{ij}(t) = w_i \cdot p_{i\to j}(t) + w_j \cdot p_{j\to i}(t), \quad \forall i \in U, j \in J, \tag{6.5}$$

where $w_i + w_j = 1$. According to the estimated values of $P_{ij}(t)$, $\forall j \in J$, task agent Ω^i then either submits a new proposal to another machine (which requires re-activation of TRAP), or dispatches task i to machine j^*, where

$$j^* = \arg\min_{j\in J}\{P_{ij}(t)\} \tag{6.6}$$

Machine agent Ω^{j^*} then receives task i, puts it in its queue, and updates the priority of the tasks waiting in the queue based on its specific dispatching policy (e.g., FIFO; shortest processing time; earliest due date; save setups). After matching the tasks and the machines, TAP proceeds to its third sub-protocol for real-time monitoring of processes, exception handling, and adaptation to potential changes.

Synchronization and Time-Out Protocol (STOP)

Generally, it is assumed that machines and similar limited resources are shared by the tasks that need their services. STOP is activated when machine agent Ω^j encounters a situation in which task i (in process) must be timed-out, returned to the queue, and preempted by another task (already in the queue). Such synchronization and time-out decisions are made by machine agents—if the queue of machine j is *not empty* and the in process task *will not be late if preempted*—under any of the following circumstances:

(1) *Excessive resource occupation*. This situation occurs when the actual process of task k is considerably *slower* than expected, resulting in excessive occupation of machine j and its resources. STOP calculates a *time-out threshold* that if exceeded by task k, its process is temporarily stopped by machine agent Ω^j, and the processing of another task is started. Accordingly, task k is timed-out if

[16]Note that the time lapse between calculating $p_{i\to j}(t)$ and $p_{j\to i}(t)$ is disregarded for simplicity of presentation and calculation.

$$\Psi_j < \frac{\breve{\theta}_{kj}(t)}{\theta_{kj}(t) - \left(t - t_k^S - t_k^Q\right)}, \quad \forall j \in J, \tag{6.7}$$

where the threshold Ψ_j is a predefined positive constant ($\Psi_j > 1$); k is the task already being processed by machine j; $\breve{\theta}_{kj}(t)$ is the remaining processing time of task k; t_k^S is the time machine j started processing task i; and t_k^Q is the amount of time task k spent in the queue after t_k^S (i.e., the previous time-out intervals of k, *if any*).

(2) *Preemption by an urgent task*. In this situation, the current task k is timed-out because there is an urgent task i in the queue that will be late if not processed right away. That is, task k is timed-out if

$$\exists i \in Q: \delta_i(t) < t + \breve{\theta}_{kj}(t), \quad \forall j \in J, \tag{6.8}$$

where $\delta_i(t)$ is the due date of task i; and Q is the queue of machine j. In such situations, task k. is preempted by task i, returned to the queue, and its process is resumed after the process of task i is completed (if no other changes take place during this time).

With the aid of TRAP, SRAP, and STOP, TAP enables distributed agents to efficiently collaborate, exchange information, and make local decisions in harmony with each other (Fig. 6.11). This example is one illustration of interaction protocols for design and administration of distributed systems. The primary challenges, however, are to enhance the intelligence of agents to be able to handle more complex and uncertain situations and to organize an efficient administrative infrastructure for the agents to create, collect and communicate information. These challenges are discussed in the next section and in Chap. 8.

Example

Recall the example of the agent-based shop floor from the previous section (Fig. 6.8). Assume that the shop processes two types of parts, P_1 and P_2; each part requires only one task; each task can be performed by any of the three machines, M_1, M_2, or M_3. Tables 6.6 and 6.7, respectively, summarize the information regarding the unassigned tasks (i.e., parts), and the machines at time $t = 0$ (current time). This example is intended to numerically explain how the matching process between the tasks and the machines is designed via TAP.

(1) *TRAP*. Five task agents are created, each responsible for a specific task i. The task agents define the characteristics of their tasks, as shown in Table 6.6, and then calculate their respective preference values using Eq. (6.2), as shown in Table 6.8. These values imply the suitability of each machine for each task; e.g., Task 1 prefers Machine 1 the most, Task 2 is indifferent between

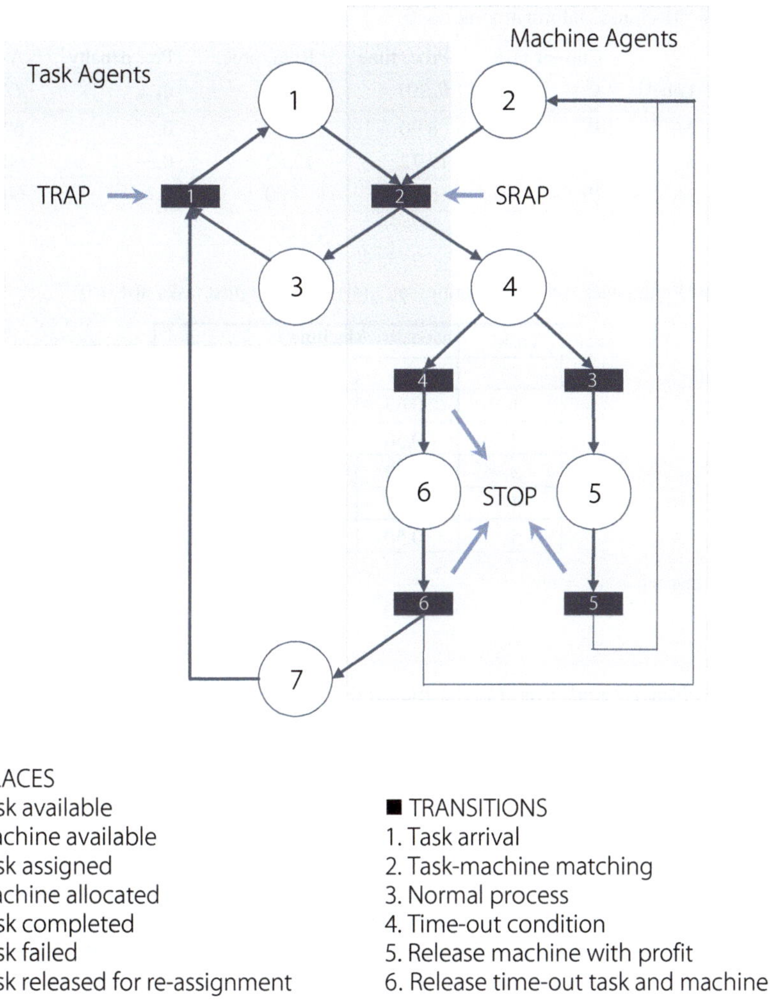

Fig. 6.11 A petri net of TAP (after Ko and Nof 2012)

Table 6.6 Unassigned tasks' it $t = 0$

Task	Type	Processing time			Due date	Profit
i	$\tau_i(0)$	$\theta_{i,1}(0)$	$\theta_{i,2}(0)$	$\theta_{i,3}(0)$	$\delta_i(0)$	$\pi_i(0)$
1	P_2	7.24	12.80	7.83	8	200
2	P_1	17.43	19.34	14.19	22	150
3	P_1	16.21	19.45	14.31	26	150
4	P_2.	6.72	12.27	7.61	8	200
5	P_1	15.68	17.41	13.67	25	150

Table 6.7 Machines' information at $t = 0$

Task		Current task	Proc. time	Rem. proc.	Pre. penalty	Av. time
j.	Label	k	$\theta_{kj}(0)$	$\breve{\theta}_{kj}(0)$	ξ_k	T_j
1	M_1	P_2	6.90	5.56	0.6	60
2	M_2	P_1	19.72	12.82	0.3	60
3	M_3	P_2	6.77	4.60	0.7	60

Table 6.8 Preferences (relative priorities; $p_{i \to j}(t)$) of unassigned tasks at $t = 0$[a]

Task	Machine j		
i	1	2	3
1	0.63	0.61	0.63
2	0.50	0.51	0.52
3	0.49	0.50	0.50
4	0.64	0.62	0.63
5	0.50	0.51	0.51

[a]Assume that $w_1 = w_2 = w_3$

Machines 2 and 3 as its best fits, and so on. The task agents then submit their information to their most preferred machine agent(s) (the shaded cells of Table 6.8), leading to the activation of SRAP.

(2) *SRAP*. Each machine agent evaluates the received proposal(s) (if any), and prepares a bid accordingly. Machine Agent 1 with a P_2 type task in process receives proposals from Task Agents 1 and 4 (both P_2 type); Machine Agent 2 with a P_1 type task in process receives proposals from Task Agents 3 and 5 (both P_1 type); and Machine Agent 3 with a P_2 type task in process receives proposals from Task Agents 1 (P_2 type), 2, 3, and 5 (P_1 type). The machine agents prepare their bid(s) (i.e., calculate their preferences) for the corresponding task agents using Eq. (6.4), as shown in Table 6.9. Subsequently, the task agents evaluate their received bid(s) using Eqs. (6.5) and (6.6), and either accept the most preferred offer or submit a proposal to another machine. Task Agent 1, for example, calculates the mutual preference values based on the bids received from Machine Agents 1 and 3 as follows:

$$P_{1,1}(0) = 0.597, \quad w_i = 0.7, w_j = 0.3,$$
$$P_{1,3}(0) = 0.588, \quad w_i = 0.3, w_j = 0.7,$$

which leads to matching Task 1 to Machine 1. This negotiation process continues until all tasks are matched.

Table 6.9 Preferences (bids; $p_{j\to i}(t)$) of machines at $t = 0$[a]

Task	Machine j		
i	1	2	3
1	0.52	–	0.57
2	–	–	0.24
3	–	0.65	0.24
4	0.52	–	–
5	–	0.65	0.24

[a]Assume that $w_1 = w_2 = w_3$; $\sum_{l\in Q} \theta_{l,1}(0) = 50$; $\sum_{l\in Q} \theta_{l,2}(0) = 45$; and $\sum_{l\in Q} \theta_{l,3}(0) = 35$

(3) *STOP*. If any of the two time-out circumstances occurs at one of the machines, the in process task is preempted by the task with the highest priority in the queue. Consider, for example, Task 1 that is already assigned to Machine 1. The processing time and due date of Task 1 are 7.24 and 8 time units, respectively (see Table 6.6); while the remaining processing time of the task already in process at Machine 1 is 5.56 time units (see Table 6.7). This example implies that Task 1 is urgent [see Eq. (6.8)]. Hence, Machine Agent 1 checks if the task in process can be preempted by Task 1 without being late, and i. so, returns it to the queue and starts processing Task 1.

6.4 The "AI" Challenges (Artificial Intelligence; Analytics and Informatics)

Distribution of decision-making abilities and authorities creates an integrated *decision-making network*, where the agents constitute the edges linked via communication channels. Such distributed architectures bring up two major cohorts of challenges corresponding to the effective localization of intelligence and autonomy, and efficient interaction and communication of information. These challenges are briefly explained in this section in the context of best matching, together with potential solutions and research directions.

6.4.1 *Artificial Intelligence*

In order for artificial agents to socialize, cooperate, act autonomously, solve problems, and make decisions, a certain level of intelligence must be programmed and embedded in them. The ultimate goal is to enable the agents to *communicate*

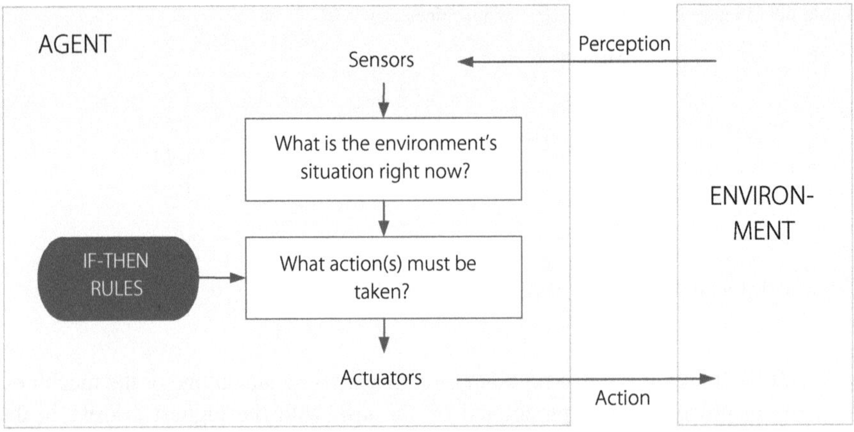

Fig. 6.12 Simple reflex agent—an agent model that functions based on basic *condition-action* (if-then) rules: if condition, then action

with their environments via sensors, and *adapt* to changes via condition-action rules, without any (or with minimum) human intervention (Fig. 6.12). In relatively simple applications such as the example of agent-based manufacturing discussed earlier in this chapter, design of agents with the capabilities of efficient communication and dynamic adaptation to changes may not be too challenging. As the size and complexity of the system scale up, however, more advanced methods must be deployed in order to guarantee the sociability, fault tolerance, autonomy, and proactivity of agents. Approaches to the augmentation of the intelligence of artificial agents are diverse, ranging from simple if-then rules to more advanced techniques such as evolutionary computation, fuzzy logic, probabilistic methods, statistical machine learning, cybernetics, and neural networks.[17] The challenges that have yet to be addressed include:

- Effective formulation, description, decomposition, and allocation of problems, and synthesis of results by distributed agents;
- Development of standard agent modeling languages for better interoperability;
- Coherence of decisions and actions to avoid potential conflicts, harmful interactions, and chaotic system behavior;
- Enabling agents to represent and reason the actions, plans, and knowledge of other agents;
- Balancing local computation, communication, and allocation of resources;
- Resolving negative impacts of problem decomposition and translation into agent-based schemas on the quality of decisions;

[17]Interested readers are referred to "Springer Handbook of Automation" (Nof, *Ed.*, 2009) and "Cambridge Handbook of Artificial Intelligence" (Frankish and Ramsey 2014).

- Development of performance evaluation and validation benchmarks for comparison with traditional systems;
- Development of self-organization mechanisms to support system evolution;
- Development of service-oriented architectures to augment the orchestration of agents and processes;
- Identification, diagnosis, recovery, and prediction of conflicts, exceptions, and disturbances.

Example

An agent-based negotiation system[18] is presented here to elaborate on the notion of artificial intelligence in multi-agent systems and its applications in best matching. In this example, the *negotiation agents* are classified into two sets of sellers and buyers with one-to-one relationships, and the goal of each agent is to match to its most preferred individual from the other set. Evolutionary computation through genetic algorithm can augment the intelligence of these agents by enabling them to learn and predict the preferences of their counterparts, and act accordingly. By negotiating on behalf of traders, intelligent negotiation agents provide various benefits, including less conflict, higher profit, lower cost of human-based negotiations, and 24/7 availability of the negotiation system. This example shows how agents can emulate the natural selection mechanisms to improve their knowledge and experience for better negotiation. The negotiation processes involve several rounds of offering/counter-offering, where each agent tries to maximize its own benefit. In each round, all agents initiate their negotiation processes by submitting offers to their counter-parts. Then, each agent evaluates the received proposals (if any), and suggests counter-offers if the received proposal is not acceptable. The *genetic negotiator* can augment this process by helping each agent predict the expectations of its owner and the counter-parts, and enhance its negotiation process as follows (Fig. 6.13):

Step 0. Define the agent as either a buyer or a seller.
Step 1. Prepare and submit proposals to counter-parts, and receive their proposals. For each received proposal, follow Steps 2–4.
Step 2. If the offer is acceptable, accept the offer and go to Step 5. Otherwise, proceed to the next step.
Step 3 Generate a population of chromosomes, each representing a candidate counter-offer for the current round of negotiation, considering the following conditions:

[18]The example is adapted from Choi et al. (2001).

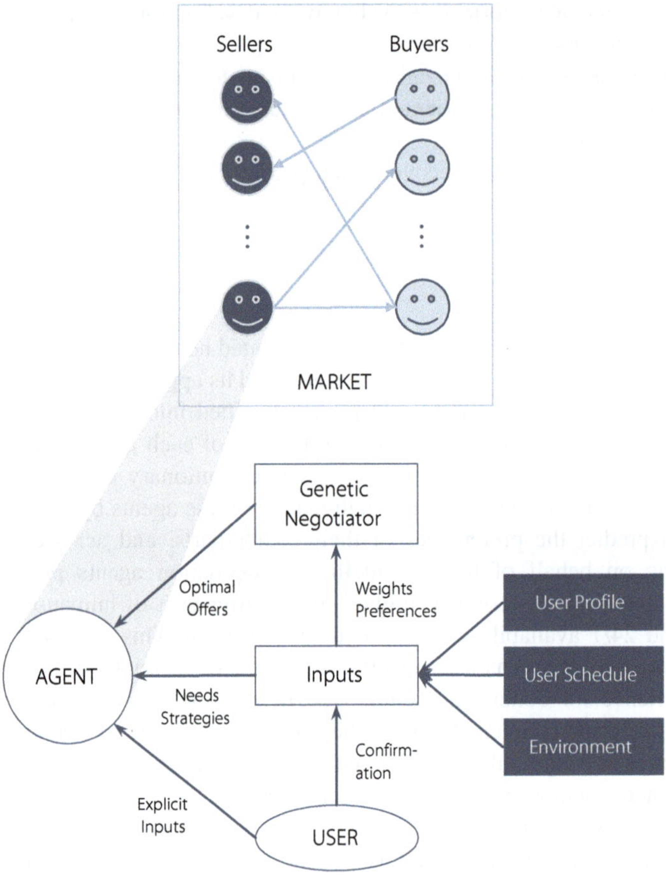

Fig. 6.13 An intelligent agent-based negotiation system based on genetic algorithm (adapted from Choi et al. 2001)

- Include a number of best previous offers and counter-offers (for convergence of the algorithm).
- None of the new offers must be better than the last offer or yield negative payoff (for feasibility).

Step 4. Reproduce a new population of chromosomes (i.e., offers) using genetic operators,[19] identify, and submit the best counter-offer. If the counter-offer is accepted, proceed to the next step. Otherwise, return to Step 3.

Step 5. Report the negotiated agreement with the best (available) match.

[19]See Chapter 5 for more information on genetic operators.

6.4.2 Analytics and Informatics

A doubtful assumption behind all the analytical models, algorithms, and protocols discussed so far is that *correct information is always available*. This is undoubtedly an untrue assumption in many practical settings, especially in emerging distributed and complex systems (e.g., networked manufacturing; intelligent transportation systems; e-Logistics; e-Healthcare) where *informatics* plays a key role in the design and control of information lifecycle and flow between different components/layers of the system (e.g., Fig. 1.2 in Chap. 1). The term informatics was coined by Karl W. Steinbuch in 1957 in his article *Informatik: Automatische Informationsverarbeitung* (in German), which means "Automatic Information Processing". Informatics is a multidisciplinary field, involving computer science, statistics, and systems engineering, and combines the science of information processing, and engineering of information systems. The ultimate goal is to assure systems designers and users that, despite the inherent complexities and interdependencies, all the distributed processing functions can be performed (near) flawlessly when converted into computer programs.

Workflow is a pragmatic approach to informatics in dynamic and complex multi-agent systems. It enables systematic design, assignment, and monitoring of tasks with high scalability, availability, and reliability. Early applications of workflow were in software engineering, involving definition of requirement plans and relationships among data processing functions in large-scale computerized software systems; e.g., financial institutions; insurance companies; education and government service systems. Workflow technology enables:

- User-friendly interfaces for editing, improving, and executing process models;
- Information exchange between heterogeneous information systems via service standardization;
- Physical transport, classification, mining, and removal of information barriers;
- Identification of documents and allocation to the best matched agent;
- Matching incoming documents to historical information;
- Deadline monitoring and resubmission, capacity exchange, and dynamic database update.

Workflows are generally developed using a standard modeling language [e.g., Unified Modeling Language (UML)], in order to systematically specify constraints, heuristics, and procedures. UML is a general-purpose modeling language that provides a visual representation of the design of a process. It contains information on (1) the path through the process, (2) the rules for following each path, (3) all the actions to be taken, and (4) the way they must be performed. With the escalated size and complexity of systems and interactions, design of workflows with *trial and error* is no longer feasible, and thus, optimal design of workflows must be included in the following phases of workflow lifecycle for the relatively more intelligent and automated processing of information (Holl et al. 2014):

(1) *Design.* This phase is related to the (re-)design of a workflow through selection and integration of components, and definition of data and the interdependencies between the components.
(2) *Development.* This phase converts the abstract design into an executable workflow by mapping the abstract elements into concrete applications, defining data sources, and sharing the designed workflow with individual agents.
(3) *Execution.* This phase maps the executables to appropriate execution environment by retrieving information on software and data sources; monitors the execution process; and delivers the execution results to the users.
(4) *Learning and feedback.* This phase evaluates and compares the actual results with the earlier projections, and repeats the cycle with recommendations on changing the parameters, choice, or execution order of components.

Example

Recall the example of selective assembly discussed in Chap. 1 (Fig. 1.1), where two sets of bolts and nuts had to be matched on a one-to-one basis. In this example, dimensional tolerance is the key criterion for customer satisfaction, and thus, best matching can play a substantial role by *aligning the imperfections* of assembly parts with each other so that the quality of final products is improved without major rework (Velásquez and Nof 2008). In this matching example, informatics is particularly important, because the parts are usually provided by widely dispersed suppliers, and the detailed inspection information gathered locally can be leveraged throughout the supply network for better quality of final products and higher customer satisfaction. A workflow architecture is presented in Fig. 6.14 for inspection activities in a distributed manufacturing environment, which was developed for a client company of the Technical Assistance Program at Purdue University in order to better meet their needs for assembly rationalization (Kang 1994). The workflow is composed of three collaborative modules: (1) Inspection station; (2) Information integration; (3) Assembly simulator. This information integration architecture enables effective information exchange and monitoring throughout distributed manufacturing sites, for better visibility of operations. The matching workflow involves four *organizational roles* including decision-maker, inspection station, matching buffer database, and e-Work station (see Fig. 6.12), which process a sequence of 20 different activities, from high-level planning to detailed control functions. This example is intended to demonstrate the differences in the design methodology, if informatics is considered along with analytical modeling, optimization, and control. Note also that matching virtually in the matching buffer database is significantly more effective than physically trying every match.

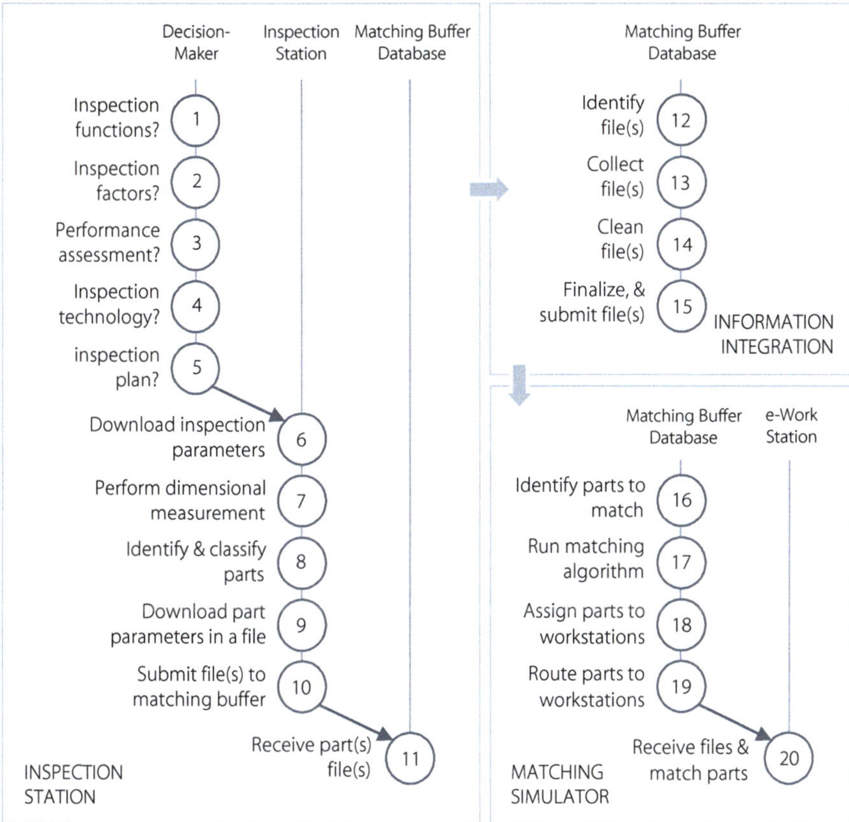

Fig. 6.14 Information exchange workflow model for collaborative multi-site inspection, matching, and assembly of parts (after Velásquez and Nof 2008)

6.5 Summary

The emerging trends towards decentralization of processes, with round the clock operations, reduced lead-times, and little/no-human decision-making, call for effective integration of analytical tools with informatics for smooth, intelligent and autonomous decision in dynamic and distributed environments. This chapter introduced the fundamentals of best matching where the D+ dimension of the PRISM taxonomy has considerable influence on the structure and characteristics of the problem, and on the availability of matching information. The transition from static to dynamic design, as discussed, can take place with either one central decision-maker or a distributed network of decision-makers (Fig. 6.15). It was shown that distributed matching (and decision-making, in general) is essentially dynamic in nature, due to the time lapse associated with communications and the

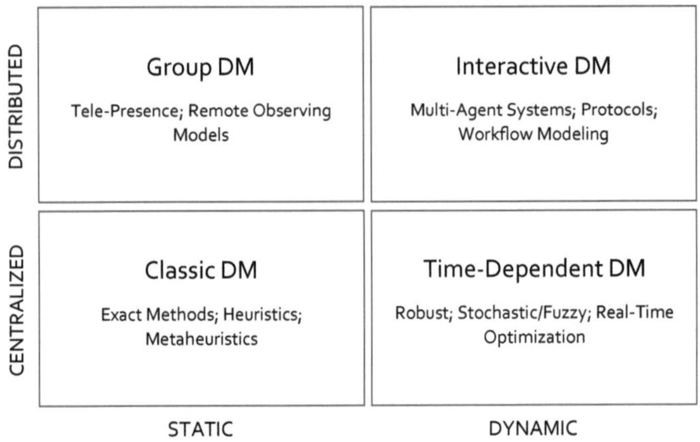

Fig. 6.15 Matching process as a function of time and number of decision-makers (DM) involved. Decision input information is assumed to be completely available in static DM; progressively available and possibly incomplete in dynamic DM

fact that the distributed decisions are not necessarily made simultaneously. It is part of the general D+ notion that best matching may be needed along progressive steps, or phases of a system throughout its lifecycle. Research on dynamic and distributed matching is still in its early stages, and several aspects of it are open for further investigation. Examples include development of adaptation heuristics for handling dynamic changes in mutual relationships/matching conditions/criteria; algorithms for detection, prevention, and resolution of errors/conflicts in matching. Chapter 8 discusses these issues along with other frontiers of matching, emerging trends, and open challenges.

Problems

6.1. List several examples of matching in manufacturing/supply/service systems with "dynamic" and "uncertain" domains. For each example,

   (a) What characteristics of the system are dynamic? Explain.
   (b) What aspects of the system are uncertain? What are the sources of this uncertainty?
   (c) According to the PRISM taxonomy of best matching, *a matching process can be specified with respect to 3 main dimensions, matching sets, conditions, and criteria, along with an additional dimension, time or progression, which addressed the "changes" in the 3 main dimensions over time or in different states.* Revise or justify your answer to (a) and (b) with respect to this statement.

6.2. What are the main limitations of the matching algorithms discussed in Chap. 5 in solving your suggested matching examples in Problem 6.1? What are your suggested approaches to address those limitations? Explain.

6.3. The word “distributed system” may refer to a system with *distributed operations* or *distributed decision-making*.

(a) What is the difference between the two aforementioned types of distributed systems?
(b) Do you agree with this classification of distributed systems? Explain why or why not.
(c) Are the algorithms discussed in Chap. 5 able to address (i) distributed operations; (ii) distributed decision-making? Explain.

6.4. List the relative advantages and limitations of the following methods in solving dynamic matching problems: (a) Robust optimization; (b) Stochastic/fuzzy optimization; (c) Real-time optimization. Is there any “overlap” between these methods? Do they complement each other?

6.5. Consider the following equations representing communication and computational cost efficiency (Λ), and system performance (Π) as functions of the length of review period (Δ).

$$\Lambda = a \cdot \log_b(\Delta),$$
$$\Pi = c \cdot \exp(d \cdot \Delta).$$

(a) Plot Λ and Π as functions of Δ for $a = 4$, $b = 1.1$, $c = 120$, and $d = -0.2$ (see Fig. 6.3).
(b) Considering the values in (a), find the optimal length of review period in $(1, 10]$.
(c) Comment on the possibility of estimating the functions and parameters associated with Λ and Π in practical settings.

6.6. Recall the example of machine, job, and operator matching from Chap. 5, Sect. 5.3.4. Assume that the system is executed under the optimal solution, and is optimized in real-time using the periodic review method with $\Delta = 2$, and at the end of the first review period (i.e., $t = 2$), the actual performances of machines M_1, M_2, and M_3. are found to be 120, 70 and 100 % of their initial (estimated) values.

(a) Using the RTO procedure with periodic review, update the optimal decision in real-time.
(b) Is your updated decision different from the initial decision?
(c) Is your updated decision at $t = 2$ subject to any changes at $t = 4$? Explain.

6.7. Recall the example of teaming for construction from Chap. 5 (Fig. 5.4). Apply the proper adaptation heuristic (RTO with continuous review) to update the optimal solution in the following cases:

(a) A new employee joins the company with the following performance and cost information:

Project	Performance (%)	Cost
1	70	$25K
2	80	$20K
3	65	$10K

(b) Employee 8 leaves the company.
(c) A new project is defined with $100K budget, where the respective performance and costs of the employees are estimated as follows:

Employee	Performance (%)	Cost
1	75	$45K
2	82	$60K
3	40	$40K
4	60	$55K
5	65	$35K
6	75	$60K
7	80	$35K
8	55	$40K
9	60	$55K
10	45	$30K

(d) Project 1 is closed.

6.8. List several advantages and limitations of the application of multi-agent systems for matching, with respect to the following performance metrics:

(a) Problem size and modularity;
(b) Timescale and changeability;
(c) Computational complexity;
(d) Rate of conflicts;
(e) Optimality of decisions.

6.9. Recall the example of teaming for construction once again (Chap. 5; Fig. 5.4). Suppose instead of central decision-making for matching, each individual employee has the authority to negotiate with the project managers regarding his/her involvement in a particular project.

(a) Develop a simple multi-agent system by assigning one agent to each individual employee and project manager. Develop an interaction (negotiation) protocol to support matching of the employees to the projects in this distributed environment.
(b) Compare your solution in (a) with the optimal solution (central) to the problem presented in Chap. 5. What changes do you notice? Which solution is better from a global (system-wide) perspective? Which solution is better from the perspective of the individuals (employee/project manager)? Why?

6.10. Using the example of intelligent negotiation system (Sect. 6.3.1), explain the benefits and drawbacks of delegating matching decisions to multiple agents, and not by an algorithm such as the examples presented in Chap. 5.
6.11. Recall the example of precision agriculture discussed in Sect. 6.2.1. Explain how TAP (Task Administration Protocols) can be used as an efficient interaction protocol for designing this system in distributed settings. You can illustrate your answer by developing a simple TAP for this example.
6.12. Suggest ways and metrics to assess the economic value of a best match obtained by a dynamic matching process, as described in this chapter, and by a static matching process (from those described in Chap. 5). Illustrate the application of your suggested methods.
6.13. Can best matching be accomplished by computer simulation (e.g., Simulink, Arena, AnyLogic, etc.)? Explain and compare to the methods described in Chaps. 5 and 6.
6.14. Best matching is essentially a process of optimization. As such, review sensitivity analysis techniques that have been developed for the methods described in this chapter, and explain how they can be useful for measuring the sensitivity of best matching solutions/decisions.

References

Bektas, T. (2006). The multiple traveling salesman problem: an overview of formulations and solution procedures. *Omega, 34*, 209–219.

Ben-Tal, A., Ghaoui, L., & Nemirovski, A. (2009). Robust optimization. Princeton: Princeton University Press.

Chachuat, B., Srinivasan, B., & Bonvin, D. (2009). Adaptation strategies for real-time optimization. *Computers & Chemical Engineering, 33*, 1557–1567.

Chituc, C. M., & Nof, S. Y. (2007). The Join/Leave/Remain (JLR) decision in collaborative networked organizations. *Computers Industrial Engineering, 53*, 173–195.

Choi, S. P. M., Liu, J., & Chan, S. P. (2001). A genetic agent-based negotiation system. *Computer Networks, 37*, 195–204.

Frankish, K., & Ramsey, W. M. (2014). The cambridge handbook of artificial intelligence. Cambridge: Cambridge University Press.

Holl, S., Zimmermann, O., Palmblad, M., Mohammed, Y., & Hofmann-Apitius, M. (2014). A new optimization phase for scientific workflow management systems. *Future Generation Computer Systems, 36*, 352–362.

Jennings, N. R., Sycara, K., & Wooldridge, M. (1998). A roadmap of agent research and development. *Autonomous Agents and Multi-Agent Systems, 1*, 7–38.

Jeong, W. (2006). Fault-tolerant timeout communication protocols for distributed micro-sensor network systems. Ph.D. Thesis (Purdue University, West Lafayette).

Jeong, W. (2009). Sensors and sensor networks. In the *Springer Handbook of Automation* (Nof, Ed.), Ch. 20.

Jeong, W., & Nof, S. Y. (2009). A collaborative sensor network middleware for automated production systems. *Computers & Industrial Engineering, 57*, 106–113.

Kang, H. (1994). Development of information exchange protocols for distributed inspection integration. M.S. Thesis. School of Industrial Engineering, Purdue University.

Ko, H. S., & Nof, S. Y. (2012). Design and application of task administration protocols for collaborative production and service systems. *International Journal of Production Economics, 135*, 177–189.

Li, J., Moghaddam, M., & Nof, S. Y. (2015). Dynamic storage assignment with product affinity and ABC classification—a case study. *International Journal of Advanced Manufacturing Technology,*. doi:10.1007/s00170-015-7806-7

Liu, Y., & Nof, S. Y. (2004). Distributed micro flow-sensor arrays and networks (DMFSN/A): Design of architectures and communication protocols. *International Journal of Production Research, 42*, 3101–3115.

Moghaddam, M., & Nof, S. Y. (2015a). Best matching with interdependent preferences—Implications for capacitated cluster formation and evolution. *Decision Support Systems, 79*, 125–137.

Moghaddam, M., & Nof, S. Y. (2015b). Real-time administration of tool sharing by best matching to enhance assembly lines balance ability and flexibility. *Mechatronics, 31*, 147–157.

Moghaddam, M., & Nof, S. Y. (2016). Real-time optimization and control mechanisms for collaborative demand and capacity sharing. *International Journal of Production Economics, 171*, 495–506.

Nof, S. Y. (2000). Intelligent collaborative agents. In C. Moore, et al. (Eds.), *Encyclopedia of science and technology* (pp. 219–222). New York: McGraw Hill.

Nof, S. Y. (2003). Design of effective e-Work: Review of models, tools, and emerging challenges. *Production Planning and Control, 14*, 681–703.

Nof, S. Y. (2007). Collaborative control theory for e-Work, e-Production, and e-Service. *Annual Reviews in Control, 31*, 281–292.

Nof, S.Y. (2009). Springer handbook of automation. Berlin: Springer.

Nof, S. Y., Ceroni, J., Jeong, W., & Moghaddam, M. (2015). Revolutionizing Collaboration through e-Work, e-Business, and e-Service. Berlin: Springer.

Pentico, D. W. (2007). Assignment problems: a golden anniversary survey. *European Journal of Operational Research, 176*, 774–793.

Pistikopoulos, E. N., Dua, V., Bozinis, N. A., Bemporad, A., & Morari, M. (2002). On-line optimization via off-line parametric optimization tools. *Computers & Chemical Engineering, 26*, 175–185.

Sahinidis, N. V. (2004). Optimization under uncertainty: State-of-the-art and opportunities. *Computers & Chemical Engineering, 28*, 971–983.

Smith, R. G. (1980). The contract net protocol: High-level communication and control in a distributed problem solver. *IEEE Transactions on Computers C-29*, 1104–1113.

Van Dyke Parunak, H. (1997). "Go to the ant": Engineering principles from natural multi-agent systems. *Annals of Operations Research, 75*, 69–101.

Velásquez, J. D., & Nof, S. Y. (2008). Integration of machine-vision inspection information for best matching of distributed components and suppliers. *Computers in Industry, 59*, 69–81.

Wald, A. (1945). Statistical decision functions which minimize the maximum risk. *The Annals of Mathematics, 46*, 265–280.

Yoon, S. Y., & Nof, S. Y. (2011). Affiliation/dissociation decision models in demand and capacity sharing collaborative network. *International Journal of Production Economics, 130*, 135–143.

Zhang, Y., Monder, D., & Fraser Forbes, J. (2002). Real-time optimization under parametric uncertainty: A probability constrained approach. *Journal of Process Control, 12*, 373–389.

Chapter 7
Extended Examples of Best Matching

Abstract This chapter presents four extended best matching case studies associated with distributed manufacturing, supply, social, and service networks. Detailed definitions, mathematical models, solution approaches, numerical analysis and discussions are presented for each case. The purpose is to demonstrate "A–Z" of modeling, optimizing, and controlling best matching processes via four comprehensive case studies.

7.1 Understanding Through Analogy

We all learn by developing connections between a new subject with what we already know. That is why a good example sometimes works much better than hours of thorough theoretical and hypothetical explanation. Thus, to further clarify the definitions, theoretical foundations, models, and methodologies introduced so far, four extended examples of best matching are presented in this chapter. Compared to the numerical examples included in the previous chapters, the extended examples enhance understanding of best matching theory and applications from multiple perspectives, through:

1. Detailed problem definition and formulation, to illustrate different dimensions of the PRISM taxonomy of best matching;
2. Development of advanced optimization and control procedures, based on the best matching algorithms and protocols described in Chaps. 5 and 6;
3. Numerical evaluation, validation, and statistical analyses.

First, the four examples are briefly introduced.

E_1. *Collaborative Supply Networks*—Better resource sharing by best matching

The first example investigates the impacts of best matching with *resource sharing* on the performance of collaborative supply networks. Demand-capacity sharing

---

The original version of this chapter was revised: For detailed information please see erratum. The erratum to this chapter is available at 10.1007/978-3-319-46070-3_9

M. Moghaddam and S.Y. Nof, *Best Matching Theory & Applications*, Automation, Collaboration, & E-Services 3, DOI 10.1007/978-3-319-46070-3_7

between individuals within a supply network are common practice, and have become attractive strategies for both competing and non-competing suppliers (e.g., airlines; test and assembly factories; outsourced maintenance/logistics providers). The goal is to maximize profit and resource utilization, enable timely delivery to customers in spite of uncertain market demands and unexpected capacity shortages, and maximize overall stability of the supply network. Demand-capacity sharing enables the suppliers with capacity shortage to utilize the excessive capacities of other suppliers in fulfilling their unfulfilled customer demand. In spite of such benefits, however, high frequency of collaboration may impose additional costs associated with transactions, negotiations, and lateral transshipment of stocks. In this context, best matching can play a key role in minimizing the collaboration costs through dynamic matching of suppliers and customers with respect to the *gap* between customers' demand and suppliers' capacity. Best matching can also improve the collaborative process by matching between the demand/capacity sharing proposals associated with different suppliers. This problem has been formulated as a mixed-integer program, and solved for both static (without considering D+) and dynamic (considering D+) scenarios. It is proven, both mathematically and experimentally, that collaborative best matching significantly improves the performance of supply networks in terms of resource utilization, stability, total profit, and service level.

E_2. ***Collaborative Assembly Lines*—Better design by best matching**

The notion of best matching of resource that are dynamically being shared in assembly line which we attempt to balance, is analyzed in this example as a classic generalized matching problem with resource constraints and precedence relationships. A collaborative assembly framework (CAF) inspired by the design principles of the collaborative control theory is introduced to enhance the *balanceability* of assembly lines. The concept of CAF relies on dynamic utilization of resources that are idle from time to time, to eliminate (or reduce) bottlenecks. This process is accomplished via two integrated mechanisms: (1) Dynamic tool sharing between fully-loaded (i.e., bottleneck) and partially-loaded workstations; (2) Dynamic matching of tasks and workstations, and partially- and fully-loaded workstations (for tool sharing). The problem is modeled as a multi-objective mixed-integer program. The three objectives are to minimize the (1) number of workstations, (2) cycle time, and (3) total collaboration cost. The CAF is proven to guarantee *relative balanceability* of assembly lines, depending on pairwise tool compatibility and tool sharing performance. This example is also studied from both static (without considering D+) and dynamic (considering D+) perspectives. For the latter scenario, a collaborative multi-agent system with sharing-matching protocols is introduced, which executes the plan generated by the optimization model, controls the process, and modifies the tool sharing decisions in real-time. A set of numerical experiments is conducted, which indicate the superiority of the CAF over the existing non-collaborative assembly approaches in terms of three performance measures, i.e., line efficiency, utilization, and balanceability.

E_3. *Clustering with Interdependent Preferences*—Better organization by best matching

This example introduces the notion of interdependent preferences and examines their impact on the quality and accuracy of generalized matching decisions. Generalized matching has a wide variety of applications, ranging from team formation and network design to scheduling, transportation, routing, production planning, facility location, allocation, and logistics. This example defines, formulates, and analyzes an important behavior associated with the generalized matching —the mutual influence of the elements of the same set on each other's preferences, if matched to the same element of another set. Such preferences are referred to as *interdependent preferences*. The problem is formulated as a binary program, and analyzed from two perspectives: Optimal cluster formation (without considering D +) and evolution (considering D+). A set of evolutionary algorithms is introduced for handling the complexity of the cluster formation problem, and enabling the network of clusters to autonomously adapt to random changes, recover, and evolve. Results from several numerical experiments indicate: (a) Significant impact of interdependent preferences on the optimality of cluster formation and evolution decisions; (b) Efficiency of the developed evolutionary algorithms in handling the problem's complexity, and the emergent behavior of matching.

E_4. *Collaborative Service Enterprises*—Better service quality by best matching

The evolution of the Internet, clouds, information and communication technologies, and collaboration sciences has transformed traditional organizations of entities (e.g., humans; machines; enterprises) to highly distributed, internetworked, and collaborative virtual (v-) Organizations. This example investigates the combined impact of collaboration and best matching on the design and administration of collaborative networked organizations (CNO). The emerging extensions of the notion of cloud computing to areas such as manufacturing, business, education, banking, and healthcare have enabled more systematic integration, harmonization, and sharing of distributed resources for processing of dynamic and diverse pool of tasks. This example contributes to the design of CNO by optimizing the location of resources (e.g., program; computer; sensor; robot) and allocation of tasks in the network, considering both physical and virtual dimensions of CNO. Collaboration is enabled through cloud and cyber-supported communication technologies for sharing resources and electronic (e-) Tasks among remote organizations, such that the overall service level, network stability, and resource utilization are optimized. The problem is formulated as a bi-objective mixed-integer program. A tabu search algorithm is also introduced with a best matching heuristic inspired by the natural *justice rule* for handling the computational complexity of the problem. Several numerical experiments illustrate, analyze, and prove the unique impact of best matching and collaborative decisions on the optimal and efficient (re)configuration of CNO.

7.2 E_1: Collaborative Supply Networks (M: $1/RC, RS/-, OS/DI$)

Dynamic market behavior and unanticipated variations in customer demand are inevitable challenges of modern supply networks. Lateral collaboration between suppliers is known as an effective strategy to alleviate such dynamicities and uncertainties. In this context, a collaborative network of suppliers (CNS) refers to a set of independent suppliers that collaborate under certain protocols such that their individual and mutual benefits are achieved. Sharing resources, information, and responsibilities—as three pillars of collaboration (Nof 2007)—is therefore the key enabler in reducing the total costs and improving the global efficiency of CNS. In practice, the suppliers have finite capacities for fulfilling their own customer orders, which may be lower/higher than the available "local" capacity of the supplier resulting in excessive inventories/backorders. Through effective collaboration, however, the suppliers can minimize the costs associated excessive inventory or backorders by sharing their excessive capacity or demand with each other. This, in turn, will minimize the overall inventory and backorder levels, leading to lower costs, both locally and at the CNS level.[1] Collaboration is proven a promising strategy for mitigating the disruption in supply or demand, and improving the long-term performance of CNS in terms of cost, utilization of resources, and service quality level. A drawback associated with such collaborative strategies, however, is the additional costs imposed to the system, mostly through negotiations, information exchange, and lateral transshipment of physical products. Thus, the *frequency* and *quantity* of collaboration must be optimized as well.

Dynamic changes in market demands are not essentially in accordance with the variations in the available corresponding capacities of suppliers. Hence, the suppliers can be classified into either of the following categories: (1) Suppliers with capacity shortage; (2) Suppliers with extra (surplus) capacity. Collaboration enables the suppliers with capacity shortage to share their unfulfilled demand with the suppliers with extra capacity. If the suppliers are not properly *matched* to the customers with respect to their demand-capacity gap, however, the resulting unnecessary sharing diminishes the impact of collaboration. Given that all suppliers belong to one (or more) unified CNS, a pragmatic solution to this problem is to dynamically match suppliers and customers so that the gaps between the local capacities and allocated demands are minimized. This solution can be done by combining two types of decisions (Fig. 7.1):

[1]Interested readers are referred to Lee et al. (2007), Kutanoglu and Mahajan (2009), Yoon and Nof (2010, 2011), Tiacci and Saetta (2011), Torabi and Moghaddam (2012), Seok and Nof (2014), Moghaddam and Nof (2014, 2016).

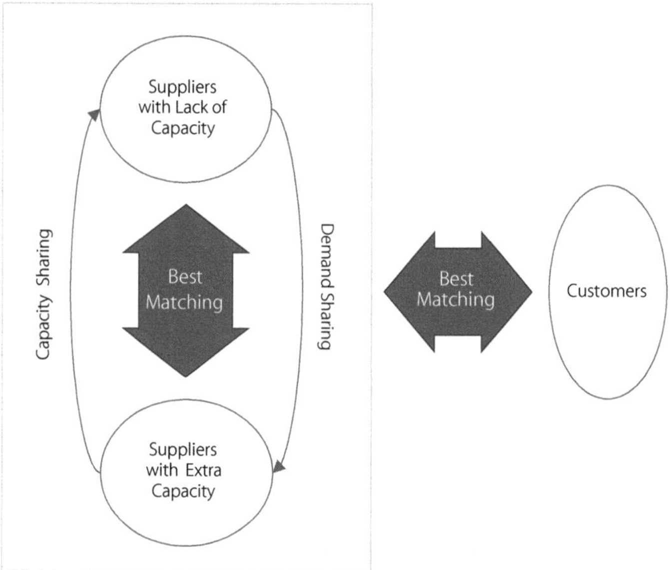

Fig. 7.1 Integration of demand and capacity sharing with best matching decisions (after Moghaddam and Nof 2014)

1. *Resource sharing*. Optimizing the overall demand fulfillment, utilization of resources, stability of CNS, and inventory holding and backordering costs;
2. *Best matching*. Optimizing resource sharing decisions—best matching between suppliers and customers, and between collaboration proposals.

The problem is addressed from two main perspectives. First, it is formulated as a mixed integer program, and solved using standard CPLEX solver of the General Algebraic Modeling System (GAMS), in order to investigate the notion of collaborative best matching in supply networks from a long-term planning (static) perspective (Sect. 7.2.1). A real-time optimization (RTO) mechanism coupled with task administration protocol (TAP) is then developed for priority-based allocation/re-allocation of resources and real-time monitoring of collaborative processes (Sect. 7.2.2). The TAP is enhanced with a predictive best matching protocol (PBMP), which optimizes allocation decisions in real-time while taking into account potential future events. The RTO mechanism is proven an advantageous solution for mitigating the undesirable impacts of uncertainty and dynamicity on the competitive performance of CNS. Figure 7.2 shows the standard specification of the problem according to the PRISM taxonomy of best matching.

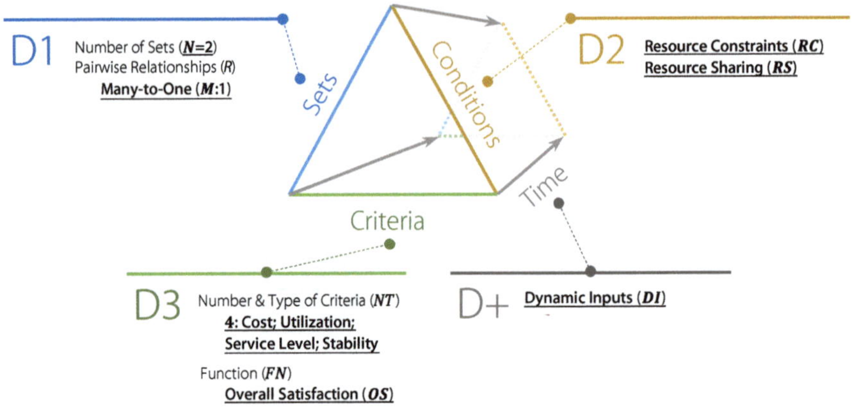

Fig. 7.2 Specific classification of the best matching with resource sharing problem in supply networks according to the PRISM taxonomy of best matching

7.2.1 Mathematical Formulation[2]

Consider a CNS where the dynamic demand of a set of customers I must be satisfied by a set of collaborative suppliers J over a planning horizon of T. Each supplier is responsible for satisfying the demand of its own customers. Depending on the dynamic variations in the customers' demands and suppliers' capacities, the customers can be served by different suppliers in different periods. In each period, after matching the customers to their respective suppliers, demand sharing (if necessary) is performed by the suppliers with capacity shortage through negotiations with the suppliers with extra capacity, and capacity sharing is accomplished through lateral transshipment of products between the suppliers. In this section, it is assumed that the matching between the collaborating suppliers is performed centrally by the developed mixed integer program. It is assumed that each supplier has finite capacity. In addition, shared fulfillment is not allowed; i.e., each customer must be served by exactly one supplier. Moreover, it is assumed that demand-capacity sharing is performed instantaneously, and backorders are allowed at all suppliers.

Following the above definitions, a framework is developed for modeling and implementation of the resource sharing and best matching decisions. The demand-capacity sharing and best matching decisions are made dynamically, taking into account the available capacities of suppliers, the demand forecasts related to each customer, and the inventory and backorder levels. The framework yields the optimal matches between the suppliers and customers, the optimal matches between the collaboration proposals, and the optimal decisions recommending production

[2]**Note**: The mathematical notations used in the four extended examples are *independent* and, in some cases, may overlap for clarity. See Appendix 1 at the end of this chapter.

and collaboration plans (see Moghaddam and Nof 2014). The problem can be formulated as the following mixed integer program[3] (see Appendix 1 for notations):

$$\text{Minimize} \sum_{j \in J} \sum_{t \in T} \left(p_{jt} Q_{jt} + h_{jt} H_{jt} + b_{jt} B_{jt} + \sum_{j' \in J} \left(v_{jj't} S_{jj't} + f_{jj't} \lambda_{jj't} \right) \right). \tag{7.1}$$

Subject to

$$\sum_{i \in I} \chi_{ijt} D_{it} = H_{j,t-1} - B_{j,t-1} + Q_{jt} - H_{jt} + B_{jt} + \sum_{\substack{j' \in J \\ j \neq j'}} \left(S_{j'jt} - S_{jj't} \right), \quad \forall j \in J, t \in T, \tag{7.2}$$

$$\sum_{j \in j} \chi_{ijt} = 1, \quad \forall i \in I, \; t \in T, \tag{7.3}$$

$$S_{jj't} \leq M \lambda_{jj't}, \quad \forall j, j' \in J, \; t \in T, \tag{7.4}$$

$$Q_{jt} \leq K_{jt}, \quad \forall j \in J, \; t \in T, \tag{7.5}$$

$$Q_{jt}, H_{jt}, B_{jt}, S_{jj't} \geq 0, \quad \chi_{ijt}, \lambda_{jj't} \in \{0, 1\}, \; \forall i \in I, j, \; j' \in J, \; t \in T. \tag{7.6}$$

Objective Function (7.1) minimizes the total cost of production, inventory holding, backordering, and lateral transshipment between the suppliers, along with fixed cost of collaboration. Constraints (7.2) guarantee inventory balance of each supplier—they imply that the possible strategies for holding the balance between capacity and demand are inventory holding/backordering, collaboration through demand-capacity sharing, and matching the suppliers' capacities with the customers' demands. Constraints (7.3) imply that each customer must be matched to exactly one supplier in each period—there is no limitation on the number of customers matched to each supplier but their available capacity. Constraints (7.4) ensure that suppliers can share their demand and capacity only if their collaboration proposals are matched, considering the fixed cost of collaboration. Constraints (7.5) guarantee that the production level of each supplier does not exceed the capacity limit. Constraints (7.6) ensure the feasibility of the decision variables. It is proven that the combined best matching-resource sharing framework significantly improves the stability of supply networks[4]:

Proposition 7.1 *Considering the entire CNS as a single "server", if the overall arrival rate* (i.e., *demand*) *is lower than the overall service rate* (i.e., *capacity*), *all suppliers are guaranteed to have stable processes through collaboration.*

---

[3]This is an extension of Model (M 3.7), $M\{:\}1/RC, RS/+, OS$, presented in Chap. 3.

[4]See Moghaddam and Nof (2014) for proof.

7.2.2 D+: Task Administration Protocol

After development of the initial plan using the aforementioned model, an effective control mechanism is required to handle potential *changes* in real-time; i.e., dynamically rationalize, coordinate, and harmonize the distributed operations (see Chap. 6). The TAP (Task Administration Protocol) enables powerful control mechanisms through three sub-protocols for priority-based task initialization, resource-aware task allocation, and task monitoring re-allocation (see Chap. 6, Sect. 6.3.2). In the context of this example, each customer order is formalized as a *task*, and the suppliers refer to the *resources* used for processing those tasks. The TAP is applied for dynamic matching of orders to resources in real-time, inspired by the notion of model predictive control; i.e., matching individuals (e.g., proposals; suppliers) in the current timeslot taking into account the characteristics of the system in the near future timeslots.[5] The RTO (Real-Time Optimization) is concerned with the total costs of collaboration (e.g., fixed costs of negotiation, information sharing) along with the variable costs of transshipment, and service level (Moghaddam and Nof 2016).

The TAP repeatedly identifies the current state of each individual supplier through distributed agents and following certain workflows, and takes necessary actions to handle the dynamic changes in the inputs (i.e., $\mathrm{D}+:=DI$; see Fig. 7.2). Task administration in this case—managing and organizing the process of receiving, processing, and delivering orders—ranges from priority-based initialization of tasks to allocation of resources and monitoring time-out conditions. The protocols are triggered for real-time control of the decisions, and are the core of the collaborative control mechanism for coherent and integrated administration and synchronization of the distributed matching and sharing processes. The ultimate goal of the TAP is to uphold the service level; i.e., reduce: late/partial fulfillment or rejection of orders, conflicts in sharing decisions (e.g., incentives to collaborate), and disruptions in demand/capacity. The three sub-protocols of the TAP, the task requirement analysis protocol (TRAP), the shared resource allocation protocol (SRAP), and the synchronization and time-out protocol (STOP), are designed as follows (Fig. 7.3):

1. *TRAP*. Each order is directed to the corresponding supplier according to the plan generated at the beginning of the period—order o_i arrives at supplier j, if $\chi_{ij}^{*}=1$, where superscript $*$ denotes optimality. Each order is specified as $o_i=\{\tau_i,\alpha_i,\delta_i\}$, where τ_i, α_i, and δ_i denote its processing time, quantity, and due date, respectively. Upon arrival of a new order, the task agent (TA) updates the priority of all orders in the queue. If all the priority values are the same, or the priority of the order is less than the other orders, it is simply added to the end of the queue, following the FIFO (First-In, First-Out) discipline. Otherwise, the queue is reordered according to the updated relative priorities. Without loss of

[5]The TAP presented here is an example of the *periodic review methods* discussed in Chap. 6.

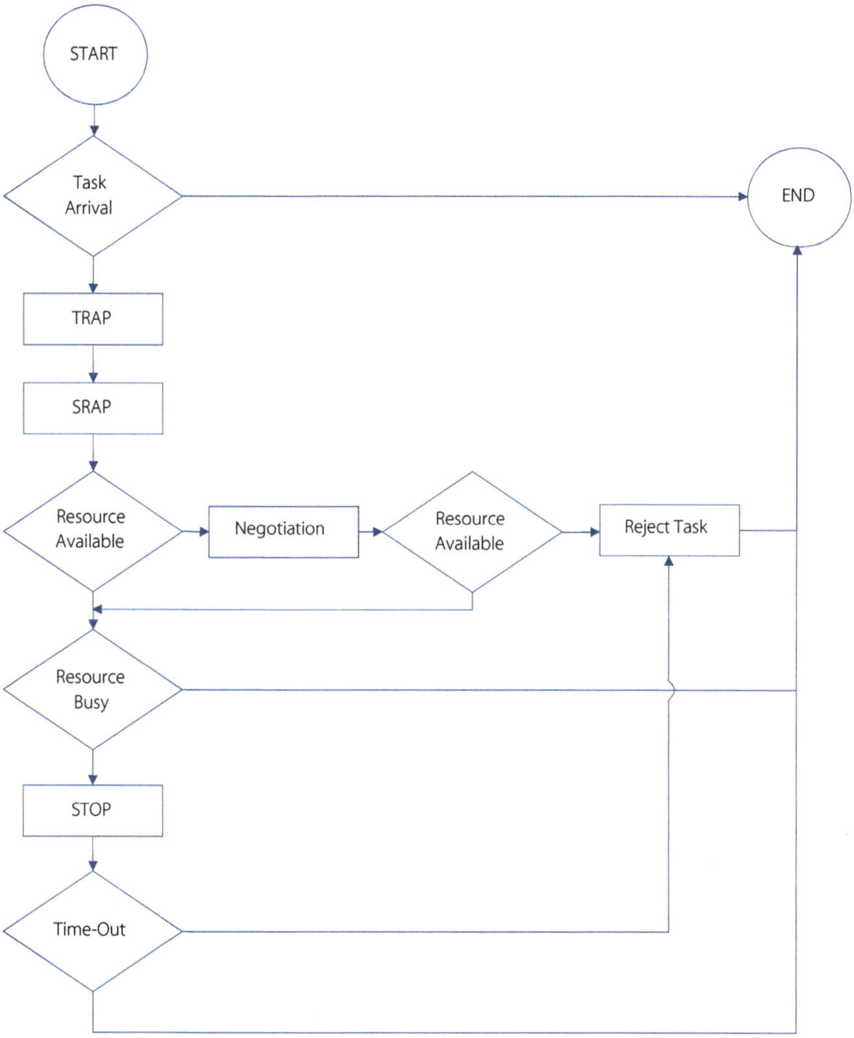

Fig. 7.3 Workflow of the TAP for distributed and dynamic matching (after Moghaddam and Nof 2016)

generality, the priorities can be calculated using the earliest due date (EDD) and shortest processing time (SPT) disciplines for lower backorders and higher throughput, respectively. Following the EDD and SPT disciplines, the priority of order o_i can be calculated through the following formula:

$$\pi(o_i, t) = w_1\left(1 - \frac{\delta_i - t}{\sum_{o_{i'}(t)} (\delta_{i'} - t)}\right) + w_2\left(1 - \frac{\tau_i}{\sum_{o_{i'}(t)} \tau_{i'}}\right), \quad (7.7)$$

where t denotes the current timeslot, $o_{i'}(t)$ denotes the orders already assigned to the resource agent (RA) at t, including the new order o_i. w_1 and w_2 denote the weights of the EDD and the SPT disciplines, respectively. After defining the priorities, each TA reports the tasks to the corresponding RA that is responsible for assigning resources to the tasks based on their relative priorities.[6]

2. *SRAP*. The RAs follow the production, collaboration, and matching decisions made by the initial plan on condition that no changes are detected in real-time. If changes take place, however, the RAs collaborate to update their initial plan using the SRAP. The RA of supplier j defines itself as a demand sharing supplier at timeslot t, if at least one of the following conditions holds (otherwise, it is a capacity sharing supplier):

$$K_j(t) - \sum_{o_{i'}(t)} \alpha_{i'} < 0, \quad \forall j \in J, \tag{7.8}$$

$$\delta_i - t - \sum_{o_{i'}(t)} \tau_{i'} < 0, \quad \exists i \in I: \chi^*_{ij} = 1, \pi(o_{i'}, t) > \pi(o_i, t). \tag{7.9}$$

In Condition (7.8), $K_j(t)$ represents situations where the available capacity of supplier j at timeslot t is not sufficient for processing its local tasks. Similarly, Condition (7.9) refers to situations where supplier j is under time limitation; i.e., there is (at least) an order o_i that will be late if processed after other orders with higher priorities. A decentralized approach for updating the sharing and matching decisions in such situations is presented below[7] (Fig. 7.4):

(a) Each RA classifies itself as "demand sharing" or "capacity sharing". If there is no demand sharing/capacity sharing RA, the procedure stops. Otherwise, it proceeds to the next step.
(b) The demand sharing RAs prepare demand-sharing proposals. The proposals include requests for partial or complete (or both) fulfillment of unsatisfied orders. Accordingly, the demand sharing RAs request *capacity sharing* for orders shared partially, and *direct delivery* from the capacity sharing suppliers in the case of complete sharing of an order.
(c) Demand sharing RA j assigns a priority value to each target capacity sharing supplier j' using the following formula:

$$\pi_{j'} = 1 - \frac{v_{jj'}}{\sum_{l \in CS_j} v_{jl}}, \tag{7.10}$$

[6]Interested readers are referred to Ko and Nof (2012) for a more comprehensive priority evaluation function.

[7]See also Moghaddam and Nof (2016) for a mediator-based approach enabled by a predictive best matching protocol.

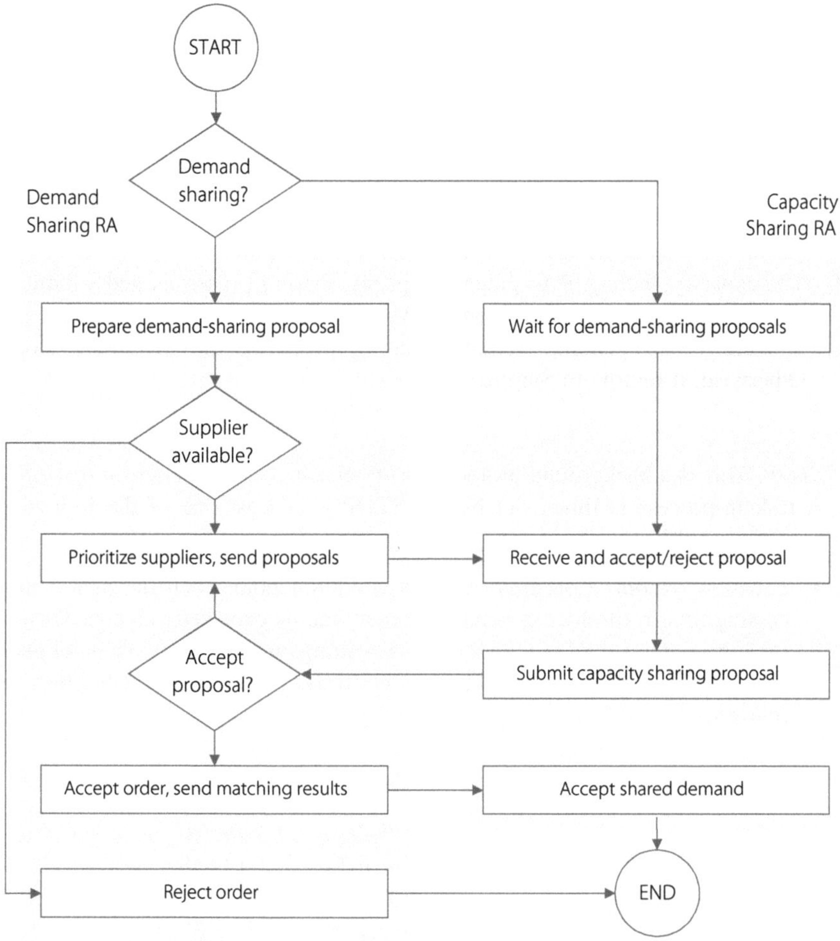

Fig. 7.4 Workflow of the decentralized collaboration procedure (after Moghaddam and Nof 2016)

where CS_j denotes the set of all target capacity sharing suppliers of supplier j. The target suppliers are sorted according to their unit transshipment costs (i.e., variable cost of collaboration). The goal is to minimize the collaboration costs.

(d) The demand sharing RAs submit their demand sharing proposals to the capacity sharing RAs following Eq. (7.10). If no capacity sharing supplier is willing to collaborate, the demand sharing RA rejects the excessive orders or keeps them as backorders.

(e) Capacity sharing RA j analyzes the received proposals (if any), and accepts order $o_{j'}$ from supplier j' if

$$\alpha_{j'} < K_j(t) - \sum_{o_i(t)} \alpha_i, \tag{7.11}$$

where $o_i(t)$ denotes the tasks that are already in the queue of supplier j. If Eq. (7.11) does not hold, order $o_{j'}$ is *partially* accepted due to capacity shortage, and the quantity of the capacity sharing proposal is then equal to the available capacity of supplier j. The demand sharing proposals are processed by the capacity sharing RAs following FIFO discipline.

(f) The capacity sharing RAs prepare capacity sharing proposals and submit to the corresponding demand sharing RAs.

(g) If there is no pending demand sharing proposal, the procedure stops. Otherwise, it returns to Step (d).

3. *STOP*. This is a background protocol for monitoring the processes in real-time. A task in-process is timed out by the STOP if at least one of the following conditions holds:

- *Excessive resource occupation*. This condition hold when the *actual* processing time $\hat{\tau}_i$ of order o_i is much higher than its expected value τ_i. This, in turn, may delay the rest of the tasks waiting in the queue. A predefined threshold is defined for checking the respective time-out condition as follows:

$$\hat{\tau}_i - \tau_i > \varepsilon, \tag{7.12}$$

where ε denotes the threshold. Nevertheless, there are two instances where the current task is not timed out even if Eq. (7.12) holds: (a) There is no other task waiting in the queue; (b) The in-process task will be delayed if timed out.

- *Preemption by urgent tasks*. This condition holds when a task may be late if not processed before the completion of the current task. Although in the TRAP prioritizes the tasks based on the EDD policy, this situation is likely to occur due to dynamic changes in the due dates and/or processing times. The preemption procedure is performed by the RAs as follows:

(a) The RA checks the following condition for all orders in the queue:

$$\delta_i - t - \tau_i < 0, \tag{7.13}$$

(b) If Condition (7.13) holds for order o_i in the queue and for order $o_{i'}$ in process (i.e., $\delta_i > \tau^r_{i'}$, where $\tau^r_{i'}$ denotes the remaining processing time of order $o_{i'}$ at timeslot t), order $o_{i'}$ is preempted by order o_i and resumed upon completion of o_i.

7.2.3 Measured Impact

The impact of combined best matching and resource sharing decisions on the performance of supply networks is significant. Experiments have investigated and analyzed such impacts from the perspective of four main criteria associated with supply network operations[8] (Fig. 7.5):

1. *Order fulfillment*. Under non-collaborative scenarios, the capacity-demand gap results in excessive inventories or backorders at the end of each period. Under the collaborative scenarios, on the other hand, demand and capacity sharing minimizes this gap through lateral collaboration. Nevertheless, since the capacities of the suppliers are limited, even the collaborative scenarios may result in certain amounts of inventories/backorders. Experiments indicate the level of unfulfilled orders is much lower when such collaborative decisions are enhanced with best matching. Some suppliers may be reluctant to collaboration due to high costs of collaboration compared to inventory holding/backordering; thus, it is recommended to eliminate "unnecessary collaborations" by minimizing the capacity-demand gap through best matching, prior to making any collaboration decisions.
2. *Resource utilization*. For each supplier, resource utilization is defined as the percentage of resources actually consumed relative to the amount of resources planned to be consumed. With this definition, resource utilization is 100 % in all demand sharing suppliers. However, collaboration can also improve the resource utilization of capacity sharing suppliers—it indeed enhances the utilization of the *existing* resources inside the CNS rather than investing in *external* resources for dealing with the local capacity shortages. That is the reason collaborative scenarios outperform non-collaborative scenarios in terms of resource utilization.
3. *Cost effectiveness*. Beside demand fulfillment and resource utilization, the collaborative scenarios also improve the total cost of the CNS compared to non-collaborative scenarios. The substantial decrease in the total cost is due to the elimination of unnecessary inventories and backorders through demand-capacity sharing. In addition, the total cost is even lower when collaboration is combined with best matching. This is the result of reducing the capacity-demand gaps of the suppliers through best matching, which results in lower rates of lateral collaboration throughout the CNS.
4. *Stability*. It is proven that if the overall demand rate is lower than the overall production rate, all suppliers are guaranteed to have stable operations under the

[8]The results presented here are the summary of the findings reported by Moghaddam and Nof (2014, 2016).

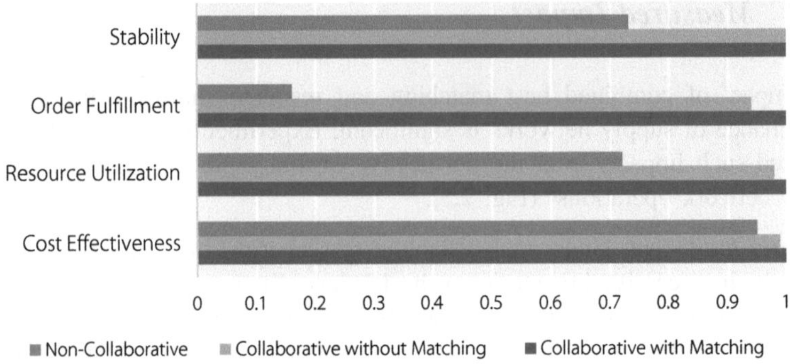

Fig. 7.5 Comparative analysis of three scenarios in terms of collaboration and best matching (Results normalized in [0, 1].)

collaborative scenarios (see Proposition 7.1). In practice, however, due to the disruptions in demand and/or capacity, CNSs are not necessarily guaranteed to undergo overall stable processes. However, collaboration increases the stability of individual supplier as much as possible. In the numerical example shown in Fig. 7.5, the overall process of the CNS is unstable or *oversaturated*. Nevertheless, results show that even in the case of overall instability, the collaborative scenarios still outperform the non-collaborative scenarios in terms of stability.

In the decentralized SRAP procedure discussed earlier, the RAs operate independently and with respect to their own local benefits. This purely heterarchical model (see Chap. 4) certainly does not guarantee the global optimality of the CNS operations in terms of total cost and service level. Experiments show that implementation of semi-heterarchical models can effectively improve the performance of the system through deployment of mediator agents (Barbati et al. 2012). A centralized predictive collaboration mechanism organized by matching agents (as mediator) is shown in Fig. 7.6. The role of the matching agent is to integrate all the distributed TAs and RAs, and control the demand-capacity sharing decisions based on the current states of all collaborating suppliers taking into account potential future events (See Moghaddam and Nof (2016) for details.). According to the experiments, the semi-heterarchical model provides significantly better solutions (confidence level of 95 %) in terms of real-time collaboration costs compared to the purely-heterarchical approach (Fig. 7.7).

7.2.4 Discussion

Collaboration is a powerful strategy for mitigating the inconsistencies between demand and capacity levels and their unforeseen variations over time. It improves

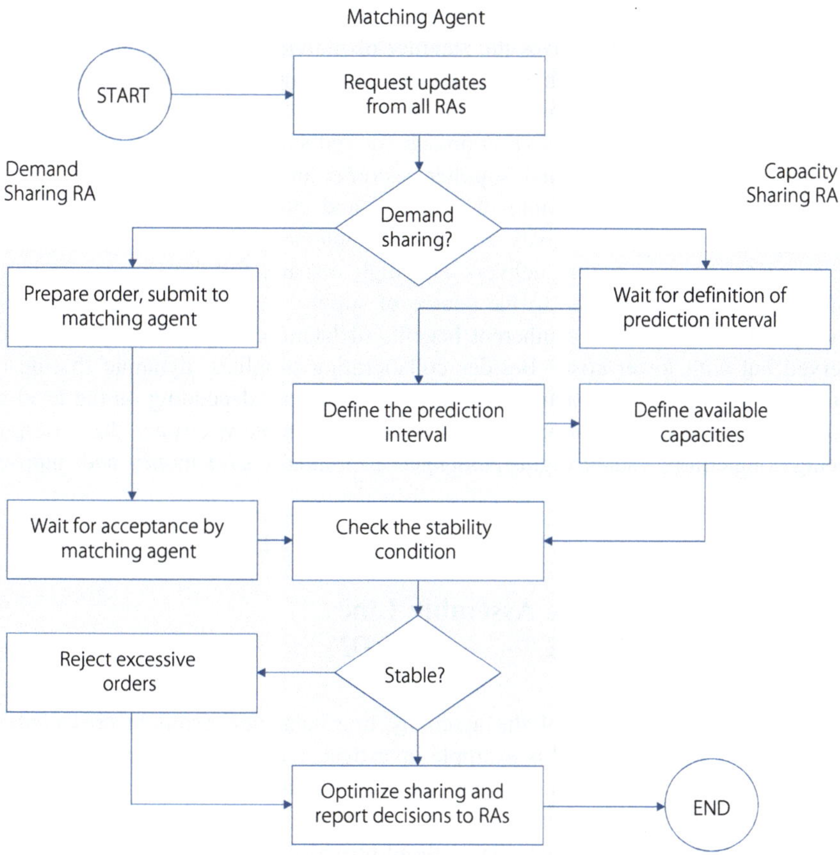

Fig. 7.6 Workflow of the semi-heterarchical collaboration procedure (after Moghaddam and Nof 2016)

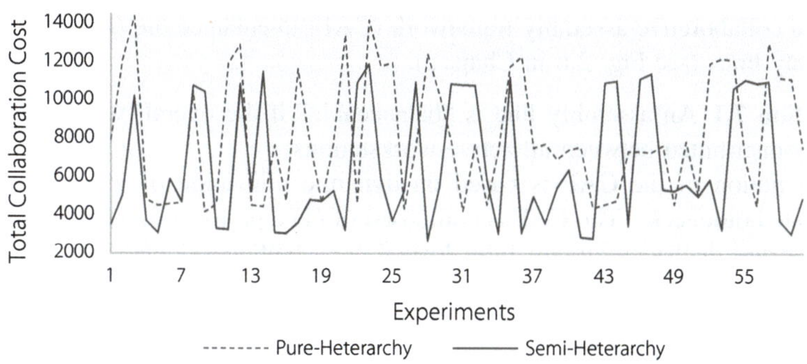

Fig. 7.7 Comparative analysis of pure-heterarchy versus semi-heterarchy in 60 independent experiments

service level, resource utilization, stock and service costs, safety stock level, and stockout rates. It also improves the stability of individual suppliers as well as the entire CNS. In spite of its benefits, collaboration imposes extra costs associated with negotiation, transactions, and transshipment of stocks. In this context, best matching is proven an effective approach for reducing the cost of collaboration through dynamic and optimal supplier-customer matching. Nevertheless, some experts perceive fixed matching of suppliers and customers as an advantageous strategy due to less complexity and easier implementation. From the customer viewpoint, however, timely delivery and high demand fulfillment rate are more important factors compared to the source of supply. Moreover, from the holistic viewpoint of the CNS, the inherent benefits of lateral collaboration are also preserved but with lower costs. Besides collaborating suppliers, dynamic sharing of customers can also be attractive to competing suppliers, depending on the level of competition and collaboration costs. A common example is airlines that—despite being competitors—share flights/passengers to mutually save money and improve their service level.

7.3 E_2: Collaborative Assembly Lines (M: $1/RC, PR, RS/---, GP/DI$)

This case is an extension of the assembly line balancing example presented in Chap. 2 (see Sect. 2.2.1). This example investigates an instance of matching with precedence relationships and resource sharing in collaborative assembly lines, where a set of tasks must be matched to a set of workstations. The insights provided by this example are complementary to those provided by the previous example: In E_1, matching (of suppliers and customers) was applied as a mechanism for improving the well-established sharing (of demand-capacity) decisions. In E_2, in contrast, the classic matching (of tasks and workstations) decisions are enhanced by integrating sharing (of "tools") mechanisms into the system. This example introduces a collaborative assembly framework (CAF) to enhance the *balanceability* of assembly lines (see Fig. 2.6 in Chap. 2).

Definition 7.1 An assembly line is "balanceable" if the overall workload can be equally distributed between all active workstations.

The notion of the CAF is based on dynamic utilization of idle resources to eliminate bottlenecks. The CAF is composed of (1) a protocol for making dynamic *tool sharing* decisions among fully loaded (i.e., bottleneck) and partially loaded workstations, and (2) a best matching protocol to dynamically matches (a) tasks and workstations, and (b) collaborating workstations. The CAF is proven to guarantee *relative* balanceability of assembly lines, depending on pairwise tool compatibility and tool sharing performance.

An assembly process consists of a sequence of tasks that cannot be subdivided and must be processed at a specific workstation. Assembly line balancing is typically performed with respect to two main criteria: (1) The number of workstations (type I); (2) The cycle time (type II). Other objectives have also been investigated in literature; e.g., line/operator efficiency; workforce cost; task duplication cost.[9] Despite all the efforts with regard to these (and similar) criteria, however, the performance of assembly lines is usually limited by one their basic characteristics—*task indivisibility*. This characteristic leads to unbalanced workloads, lower flexibility and throughput, even in optimally balanced lines. In practice, fully balanceable assembly lines are difficult (or sometimes impossible) to achieve. Accordingly, workstations are typically classified as (A) fully loaded, with workloads equal to the cycle time, or (B) partially loaded, with workloads lower than the cycle time. Although Set A is essentially non-empty (i.e., there is at least one bottleneck workstation), Set B may be empty, but with a low likelihood and in case the line is fully balanceable (i.e., no further improvement possible).

In classic assembly lines, workstations are isolated and operate independently. In such settings, increasing the line throughput requires improving the performance of the bottleneck workstation(s). Nevertheless, this may not be the best strategy due to the requirement for additional investments (e.g., equipment duplication) while a part of the already existing equipment is idle in the partially-loaded workstation(s). Thus, the CAF improves the utilization of *available* idle resources rather than investing in additional resources for higher line balanceability. This is done through two major components: tool sharing and best matching. A practical example of tool sharing in assembly and test utilities is sharing of computer integrated testers and inspection tool, especially when their actual use is only during a small fraction of the assembly process at each assembly station. In the context of the CAF, a "tool" refers to any type of resources (e.g., equipment; robot; operator) that are capable of processing the tasks and can be shared between the workstations.

The matching process involves finding the best correspondence between the sets of tasks and workstations, taking into account tool sharing alternatives, processing times, precedence relationships, and cycle time upper bound. Tool sharing process is triggered upon identification of idle workstations, and involves negotiations between idle and busy workstations. Based on the tool sharing proposals received from Set B (i.e., partially loaded) workstations, the best matching module matches the proposal to the best workstation in Set A, considering tool-task compatibility and pairwise linear distance between the corresponding workstations. The problem is formulated as a multi-objective mixed-integer program, with the objective of improving the line balanceability through minimizing the number of workstations, minimizing the cycle time, and minimizing the total cost of collaboration. A CMAS (Collaborative Multi-Agent System) is also developed to enhance the automation of the CAF. Intelligent and autonomous agents, distributed among the assembly system, operate in accordance with a tool sharing-best matching plans. The real-time, collaborative control mechanism is adapted and extended to provide

[9]Interested readers are referred to Ghosh and Gagnon (1989), Becker and Scholl (2006), Boysen et al. (2007, 2008).

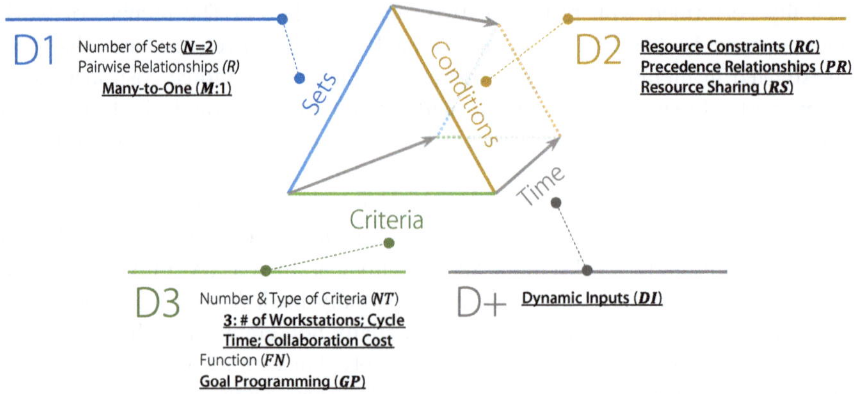

Fig. 7.8 Classification of the best matching problem with precedence relationships and resource sharing in collaborative assembly lines according to the PRISM taxonomy of best matching

feedback to the off-line plan generated and updated continuously using the optimization model. Experiments show the applicability of the CAF in significantly improving the performance of assembly lines in terms of flexibility (i.e., cycle time reduction), utilization of tools, and balanceability in real-time; i.e., considering D+ of the PRISM taxonomy (Fig. 7.8).

7.3.1 *Mathematical Formulation*

Consider an assembly process with dynamic demand of D_t ($t \in T$) over a $|T|$-period horizon. The assembly process is decomposed into a number of tasks with certain processing times and precedence relationships. Due to technological and economic limitations, an upper bound is defined for the number of workstations. An upper bound is also defined for the cycle time with respect to the demand rate and the available production time A_t as follows:

$$\overline{C}_t = \frac{A_t}{D_t}, \quad \forall t \in T. \tag{7.14}$$

The theoretical lower bound for the number of workstations is then defined according to the overall workload and the cycle time upper bound as follows:

$$\underline{W}_t = \left\lceil \frac{\sum_{j \in J} \rho_{jt}}{\overline{C}_t} \right\rceil, \quad \forall t \in T. \tag{7.15}$$

The CAF assumes that: (1) The processing times are independent from the matched workstations and are fixed in all periods. (2) The tasks cannot be subdivided between multiple workstations. (3) The idle workstations are allowed to share tools with the busy workstations. (4) No work-in-process inventory is allowed. The CAF finds the best tool sharing-best matching decisions through the following logic (Fig. 7.9):

1. The input data is collected for demand forecasts, processing times, precedence relationships, cycle time upper bound, upper bound and lower bounds for the number of workstations, available production time, and pairwise collaboration efficiency (CE) between the workstations. CE, denoted by $e_{ii'} \in [0, 1]$, represents the efficiency of the tools of workstation i in processing the tasks of workstation i' through tool sharing.

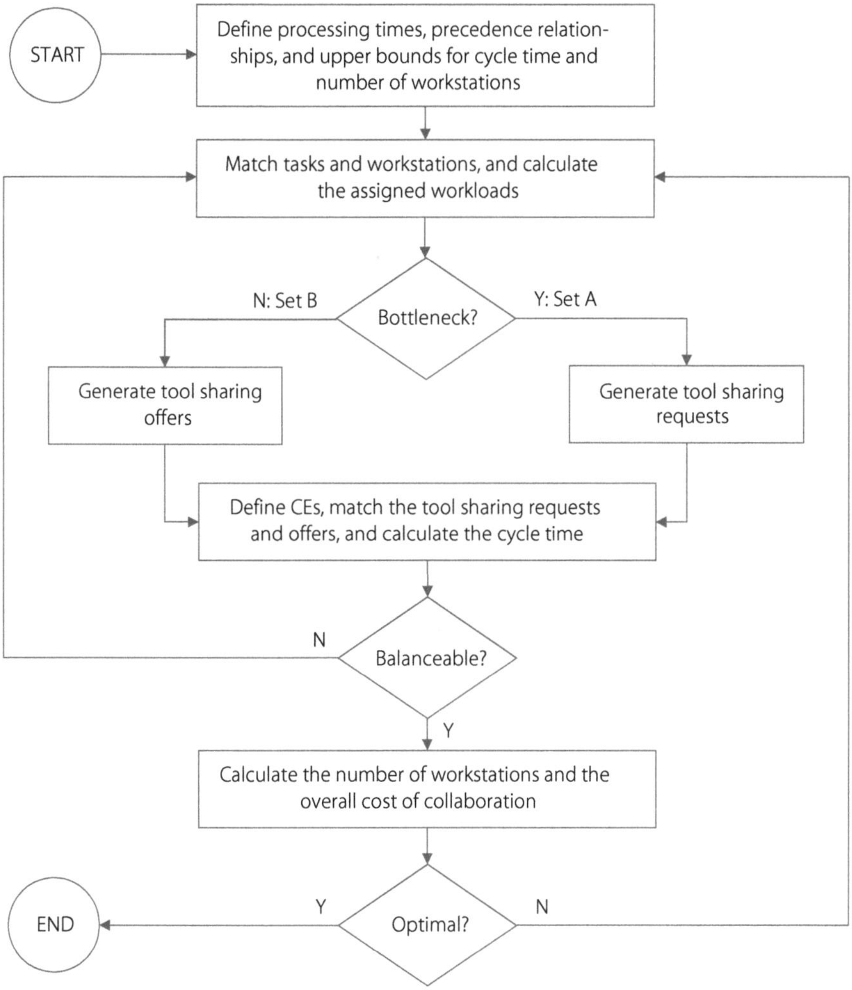

Fig. 7.9 The CAF logic (after Moghaddam and Nof 2015a)

2. The best matching protocol optimally matches tasks and workstations, considering processing times, precedence relationships, cycle time upper bound, and the lower bound on the number of workstations. This is a many-to-one matching, implying that each task must be matched to only one workstation, but each workstation may process more than one task.
3. The tool sharing protocol classifies the workstations into Sets A and B, identifies the tool sharing offers, and calculates the potential tool sharing benefits according to the pairwise CEs.
4. The best matching protocol matches the tool sharing proposals with respect to the pairwise CEs. This is a many-to-many matching, implying that a busy workstation can be assisted by more than one idle workstation, and vice versa.

This problem can be formulated as the following multi-objective mixed-integer program[10]:

$$\text{Minimize} \quad Z_1 = \sum_{i \in I} \sum_{j \in J} \sum_{t \in T} i\chi_{jit}, \tag{7.16}$$

$$\text{Minimize} \quad Z_2 = \sum_{t \in T} C_t, \tag{7.17}$$

$$\text{Minimize} \quad Z_3 = \sum_{i \in I} \sum_{i' \in I} \sum_{t \in T} f_{ii'} \psi_{ii't}, \tag{7.18}$$

Subject to

$$\sum_{j \in J} \rho_{jt} \chi_{jit} \leq \overline{C}_t, \quad \forall i \in I, \ t \in T, \tag{7.19}$$

$$\sum_{i \in I} i\chi_{j'it} \leq \sum_{i \in I} i\chi_{jit}, \quad \forall j \in J, \ j' \in Pd_j, \ t \in T, \tag{7.20}$$

$$\sum_{i \in I} \chi_{jit} = 1, \quad \forall j \in J, \ t \in T, \tag{7.21}$$

$$S_{ii't} \leq M \psi_{ii't}, \quad \forall i, \ i' \in I, \ t \in T, \tag{7.22}$$

$$\sum_{j \in J} \rho_{jt} \chi_{jit} + \sum_{i' \in I} (S_{ii't} - e_{i'i} S_{i'it}) \leq C_t, \quad \forall i \in I, \ t \in T, \tag{7.23}$$

$$\sum_{i' \in I} S_{ii't} \leq \xi \sum_{j \in J} \chi_{jit}, \quad \forall i \in I, \ t \in T, \tag{7.24a}$$

[10]This model is an extension of Model (M 3.6), M: $1/RC, PR/+, OS$, and Model (M 3.7), M: $1/RC, RS/+, OS$, presented in Chap. 3.

$$\sum_{i' \in I} S_{i'it} \leq \xi \sum_{j \in J} \chi_{jit}, \quad \forall i \in I, \ t \in T, \tag{7.24b}$$

$$S_{ii't}, C_t \geq 0, \quad \psi_{ii't}, \ \chi_{jit} \in [0, 1], \ \forall i, \ i' \in I, \ t \in T. \tag{7.25}$$

Objective Function (7.16) minimizes the number of workstations (i.e., $W_t^* = \max_{i,j}\{i\,\chi_{jit}^*\}$, $\forall t \in T$), Objective Function (7.17) minimizes the cycle time, and Objective Function (7.18) minimizes the total collaboration cost, over the planning horizon. Optimization of these conflicting objectives can be done in different ways.[11] Constraints (7.19) ensure that the workload of each workstation does not exceed the cycle time upper bound. Constraints (7.20) addresses the pairwise precedence relationships between the tasks. Constraints (7.21) ensure that each task is matched to exactly one workstation in each period. Constraints (7.22) imply that tool sharing must be performed between workstations i (idle) and i' (busy), if their collaboration proposal is approved with respect to their pairwise collaboration cost and CE. Constraints (7.23) define tool sharing variables (i.e., $S_{ii't}$) as "virtual extra times" that the idle workstations share with the busy workstations to reduce their workload. Since the conflicting objectives of minimizing the number of workstations and the cycle time are considered simultaneously, the model sometimes generates some "redundant" workstations without any allocated tasks, just for the sake of tool sharing. To avoid such conflicts, Constraints (7.24) are defined which only allow active workstations to share tools. Constraints (7.25) ensure the feasibility of the decision variables.

Following the definition of balanceability, the necessary condition for a non-collaborative assembly line to be balanceable is $\sum_{j \in J} \rho_{jt}\chi_{jit} = C_t^B$, $\forall t \in T$, where C_t^B denotes the cycle time of an equivalent balanceable line. The CAF is proven to guarantee the balanceability of assembly lines when all tools are 100% compatible and the pairwise distances between the workstations are negligible[12].

Proposition 7.2 *The CAF guarantees balanceability of an assembly line if* $e_{ii'} = 1$, $\forall i, i' \in I$.

Even if $\exists\, i, i' \in I$, $e_{ii'} < 1$, *the CAF still yields a* relatively *balanceable assembly line, where the overall workload is "almost" equally distributed between the workstations, and the cycle is always higher than the cycle time of an equivalent fully balanceable linetime*[2]:

Proposition 7.3 *If* $\exists\, i, i' \in I$, $e_{ii'} < 1$, *the CAF leads to "relative balanceability" of an assembly line, where the lower bound for the gap between the cycle time and* C_t^B *is equal to* $\delta_t = \sum_{i \in I} \sum_{i' \in I} (S_{ii't} - e_{i'i}S_{i'it})/W_t^*$, $\forall t \in T$.

[11] A *goal programming* approach is proposed by Moghaddam and Nof (2015a) for this formulation, which is not presented here for the sake of brevity.

[12] See Moghaddam and Nof (2015a) for proof.

7.3.2 D+: Collaborative Multi-agent System

A CMAS (Collaborative Multi-Agent System) is developed for real-time execution, monitoring, and modification of the CAF plan generated in the previous section. The CMAS is composed of a loosely coupled network of heterogeneous agents that collaborate to improve the flexibility and balanceability of assembly lines. Two types of agents are deployed. Workstation agents (WAs) are responsible for monitoring the processes and calculating the estimated workload in each cycle. The WAs can be equipped with simple vision sensors and processors for calculation of the assembly task progress in real-time. Tool agents (TAs) are responsible for negotiations regarding collaboration with other workstations. The architecture of the integrated planning-control mechanism developed here is semi-heterarchical with a supervisory control. Accordingly, despite the heterarchy between the agents (i.e., WAs; TAs), certain activities and responsibilities are periodically modified and updated by a mediator agent (Fig. 7.10).

The plan is generated by solving the CAF optimization model and obtaining the optimal values for the decision variables. Similar to the previous example, the CMAS presented here follows the periodic review method presented in Chap. 6, where during each review period Δ, the tool sharing-best matching protocol monitors and updates the plan (if necessary) as described below.

1. After becoming idle at time t, the TA of the idle workstation, denoted by k, estimates its "progress rate" as follows (Note: The current time is set to zero at the beginning of each review period):

$$\alpha_k(t) = \frac{\sum_{j \in J} \rho_j \chi_{jk}^*}{t} \tag{7.26}$$

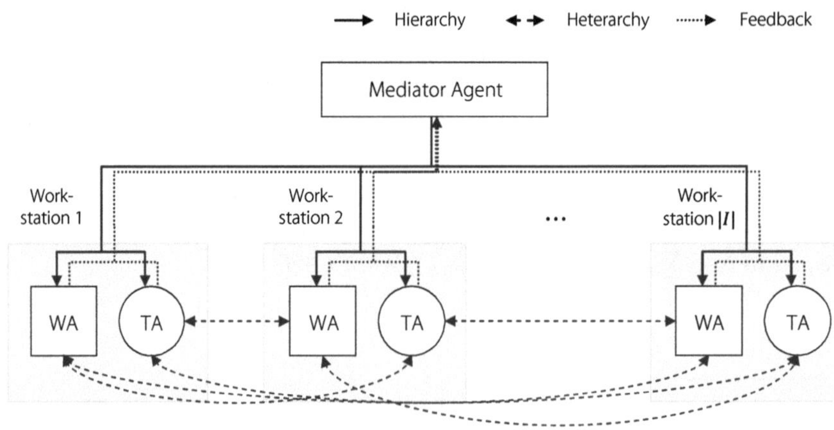

Fig. 7.10 Semi-heterarchical architecture of the CMAS (after Moghaddam and Nof 2015b)

and sends a message to the WAs of other workstations requesting their progress rates. (Superscript * denotes optimality.)

2. The WAs that received the request calculate and report their progress rates as follows:

$$\alpha_i(t) = \frac{\varepsilon_i(t) \sum_{j \in J} \rho_j \chi^*_{ji}}{t}, \quad \forall i \in I \backslash \{k\}, \tag{7.27}$$

where $\varepsilon_i(t)$ denotes the fraction of the workload of workstation i that is successfully processed by time t.

3. After receiving the progress rate reports, the TA of workstation k estimates the overall workload of all workstations as follows:

$$w_i(t) = \frac{1}{\alpha_i(t)} \sum_{j \in J} \rho_j \chi^*_{jk} + \sum_{i' \in I} \left(\frac{S^*_{ii'}}{\alpha_i(t)} - \frac{e_{i'i} S^*_{i'i}}{\alpha_{i'}(t)} \right), \quad \forall i \in I, \tag{7.28}$$

4. The TA of workstation k identifies the set of workstations that (a) are not in its predefined tool sharing plan, and (b) their overall workload is higher than the overall workload of all workstations that are in its predefined sharing plan, i.e.,

$$O_k(t) = \left\{ i{:}S^*_{ki} = 0, \quad w_i(t) > w_{i'}(t), \ \forall i' \in X_k \right\}, \tag{7.29}$$

where $X_k = \left\{ i{:}\ S^*_{ki} > 0 \right\}$. Then,

(a) If $O_k(t) = \emptyset$ and $\max_i\{w_i(t)\} \neq w_k(t)$, or $O_k(t) = \emptyset$ and $w_i(t) = w_{i'}(t)$, the TA of workstation k follows its predetermined plan.

(b) If $O_k(t) = \emptyset$ and $\exists\, i \in I$, $\max_i\{w_i(t)\} = w_k(t) > w_i(t)$, the TA of workstation k revises its tool sharing plans and levels its workload by solving the following linear program:

$$\begin{aligned} &\text{Minimize} && C \\ &\text{Subject to} && w_i(t) + e_{ki} S_{ki} \leq C, \quad \forall i \in X_k, \\ & && w_k(t) - \sum_{i \in X_k} S_{ki} \leq C, \quad \forall i \in X_k, \\ & && w_i(t) \leq C, \quad \forall i \in I \backslash \{X_k \cup \{k\}\}, \\ & && S_{ki}, C \geq 0, \quad \forall i \in X_k. \end{aligned} \tag{M7.1}$$

(c) If $O_k(t) \neq \emptyset$, the TA of workstation k revises the plan to *smoothen* its workload through solving the following linear program:

$$\begin{aligned} &\text{Minimize} \quad C \\ &\text{Subject to} \quad w_i(t) + e_{ki}S_{ki} \le C, \quad \forall i \in X_k, \\ &\qquad w_k(t) - \sum_{i \in X_k} S_{ki} \le C, \quad \forall i \in O_k(t), \\ &\qquad w_i(t) \le C, \quad \forall i \in I \backslash \{X_k \cup O_k(t) \cup \{k\}\}, \\ &\qquad S_{ki}, C \ge 0, \quad \forall i \in X_k \cup O_k(t). \end{aligned} \tag{M7.2}$$

and updates $X_k \leftarrow X_k \cup \{i{:}\ i \in O_k(t), \hat{S}_{ki} > 0\}$, where superscript ^ denotes the revised decision value obtained from Model (M 7.2).

5. The procedure returns to Step 1, and is repeated until $t = C$.
6. At the end of the cycle, the real-time performances are recorded as feedback for the next generations of the off-line plan developed by the CAF optimization model.

Control and modification of the plan in real-time are based on the "relative performance" of different workstations, represented by the progress rates $\alpha_i(t)$, $\forall i \in I$, where $\alpha_i(t) < 1$, $\alpha_i(t) > 1$, and $\alpha_i(t) = 1$ respectively imply underachievement, overachievement, and complete achievement of the expected values. Thus, if the progress rates of all workstations are similar, regardless of their corresponding values, there is no need to modify the generated plan. Otherwise, the plan must be modified such that the workstations with higher overall workload receive extra help from the other workstations in order to improve the cycle time. The modifications are performed by the TAs after their corresponding tools become idle. To better understand the above tool sharing-best matching procedure, consider the following numerical example.[13] Suppose the off-line plan is as shown in Table 7.1, with the optimal number of workstations equal to 6 and the optimal cycle time of 9.

Now consider, for example, three potential scenarios (see Step 4 of the procedure) that may occur after Workstation 4 becomes idle[14]:

1. Workstation 4 becomes idle at $t = 5$. Suppose the progress rates and actual workloads of other workstations are as shown in Table 7.2. According to Eq. (7.29), $O_4(5) = \emptyset$, and since $\max_i\{w_i(5)\} \ne w_4(5)$ (see Step 4.a), the TA of Workstation 4 follows its predetermined plan, and shares its tools with Workstation 5, since the actual performance of Workstation 5 is the lowest among all workstations. (Note that in this case, the TAs of other workstations may alter their plans after becoming idle, if the progress rates remain the same.)
2. Workstation 4 becomes idle at $t = 6$. Suppose the progress rates and actual workloads of other workstations are as shown in Table 7.3. According to Eq. (7.29), $O_4(6) = \emptyset$, and $\max_i\{w_i(6)\} = w_4(6)$, which implies that the tools of Workstation 4 may create a bottleneck in another workstation (due to underachievement of its actual progress rate), if their tool sharing plan remains

[13]The example is adopted and modified from Moghaddam and Nof (2015b).

[14]Note that in all three scenarios, the TA of Workstation 4 follows Steps 1–3 to estimate the progress rates.

Table 7.1 The off-line plan generated by the CAF optimization model

Workstation	Workload (min)	Tool sharing	
		$\sum_{i'} S^*_{ii'}$	$\sum_{i'} S^*_{i'i}$
1	7	2 $(i' = 3)$	–
2	6	3 $(i' = 3)$	–
3	14	–	5 $(i' = 1, 2)$
4	5	4 $(i' = 5)$	–
5	14	–	5 $(i' = 4, 6)$
6	8	1 $(i' = 5)$	–

Table 7.2 Scenario 1—progress rates and actual workloads at $t = 5$

Workstation	f_i	$\alpha_i(5)$	$w_i(5)$
1	0.66	0.92	9.78
2	0.92	1.10	8.18
3	0.46	1.30	5.87
4	1.00	1.00	9.00
5	0.29	0.80	12.50
6	0.63	1.00	9.00

Table 7.3 Scenario 2—progress rates and actual workloads at $t = 6$

Workstation	f_i	$\alpha_i(6)$	$w_i(6)$
1	0.79	0.92	9.78
2	1.10	1.10	8.18
3	0.56	1.30	5.87
4	1.00	0.83	10.84
5	0.48	1.12	6.68
6	0.75	1.00	9.00

the same. Thus, the TA of Workstation 4 distributes a portion of its tool sharing plan between the TAs of other workstations to smoothen its workload and minimize the actual cycle time (see Step 4.b). The TA of Workstation 4 then updates its plan by solving Linear Program (M 7.1).

3. Workstation 4 becomes idle at $t = 4$. Suppose the progress rates and actual workloads of other workstations are as shown in Table 7.4. According to Eq. (7.29), $O_4(4) \neq \emptyset$, which implies that at least one workstation (in this case, Workstation 3) that is not in the tool sharing plan of Workstation 4 is causing a bottleneck due to underachievement of its actual progress rate. Thus, in order to smoothens the overall workload and minimize the actual cycle time (see Step 4. c), the TA of Workstation 4 revises its tool sharing plan by solving Linear Program (M 7.2).

Table 7.4 Scenario 3—progress rates and actual workloads at $t = 4$

Workstation	f_i	$\alpha_i(4)$	$w_i(4)$
1	0.57	1.00	9.00
2	0.73	1.10	8.18
3	0.24	0.85	11.74
4	1.00	1.25	7.20
5	0.32	1.12	8.30
6	0.50	1.00	9.00

7.3.3 Measured Impact

The CAF has significant impact on the performance of assembly lines in terms of number of workstations, cycle time, and collaboration cost.[15] Experiments show that the CAF leads to smoother workloads compared to traditional non-collaborative approaches. It is shown that the CAF is capable of reducing the standard deviation of the workloads of different workstations down to 3 % in small-sized assembly lines and to 16 % in larger instances (Moghaddam and Nof 2015a). This is a remarkable finding that can lead to significantly better resource utilization and line efficiency, as shown in Fig. 7.11. Figure 7.12 compares the cycle time values obtained from static (without considering D+) and dynamic (considering D+) CAF. Results show that the CMAS significantly improves the performance of assembly systems through modification of the tool sharing plan with respect to the dynamic variations in the system's characteristics. The experiments also show that the gap between the results of static versus dynamic CAF is proportional to the cycle time. That is, as the cycle time of the system increases, the impact of high-/low-performance workstations on the gap between the cycle times becomes more evident. The average improvements range from 9 % for small-sized assembly lines to 25 % for larger instances (Moghaddam and Nof 2015b). This phenomenon is also intensified further, as the number of workstations increases resulting in a more diverse population of high-/low-performance workstations.

7.3.4 Discussion

The improved balanceability of assembly lines, where, for fixed line configurations, the deviation of the optimal cycle time from its ideal value (i.e., the cycle time of an equivalent balanceable line) is a function of the CE coefficients. That is, in ideal assembly line with "all-purpose" tools located in different workstations along with negligible tool transfer times, the CAF leads to fully balanceable assembly processes. Improved balanceability, in turn, leads to higher flexibility of the assembly

[15]These results presented here are the summary of the findings reported by Moghaddam and Nof (2015a, b).

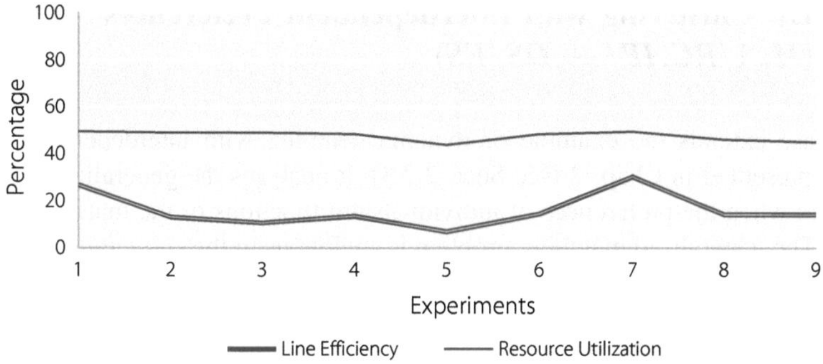

Fig. 7.11 The gap between line efficiency and resource utilization of the collaborative and non-collaborative scenarios in nine independent experiments

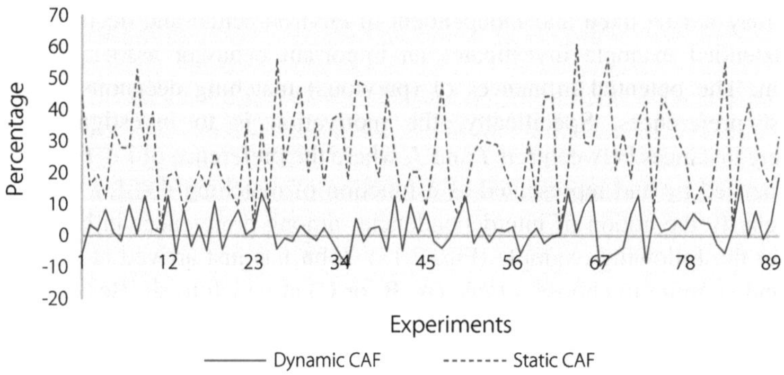

Fig. 7.12 Deviation from the planned cycle time—static versus dynamic CAF

line in dealing with dynamic variations in the market demand. Collaboration enhanced with best matching increases the maximum capacity of the assembly line in terms of throughput. The unique advantage of the CAF is the emphasis on optimal identification, assignment, and utilization of the *existing* idle resources in elimination of bottleneck workstations, rather than investing in extra resources (e.g., equipment duplication). The CMAS provides a simple yet practical example for distributed and real-time control of collaborative assembly processes. The CE coefficients are defined as the representatives of tool sharing success between each pair of workstations, considering tool-task compatibility, setup times, movement times, and so on. Case-specific technical and technological limitations may prevent the CEs from taking their highest values, and thus the assembly line from being fully balanceable. Flexibility and capability of the shared tools in joining the process of a task at any situation, and the possibility of concurrent processing of tasks with multiple tools are among the challenges of the CAF that have yet to be addressed.

7.4 E_3: Clustering with Interdependent Preferences (M: $1/RC, IP/+, OS/ES$)

This case extends the example of dynamic teaming with interdependent preferences, presented in Chap. 2 (see Sect. 2.2.3). It analyzes the generalized matching problem when the preferences of individuals are functions of the matching of their peers. The generalized matching problem is analogous to the *capacitated clustering problem*—the problem of partitioning a number of individuals into disjoint clusters with certain capacities. For example, customers, tasks, and interns may represent the sets of individuals to be respectively matched to suppliers, machines, and hospitals, each representing a specific cluster with limited capacity (e.g., a supplier can serve a limited number of customers; a machine can process a limited number of tasks; a hospital can admit a limited number of interns). In this context, clustering is performed with respect to the mutual preferences of individuals and clusters, which may not be fixed and independent of environmental and decisional factors. This extended example investigates an important behavior associated with this problem: The potential influences of (previous) matching decisions on the individuals' preferences. Specifically, the motivation is to investigate two-sided matching instances between sets I and J, where the preference of $i \in I$ for $j \in J$ may be influenced by and represented as a function of matching $i' \in I$, $i' \neq i$, to $j \in J$.

To clarify the notion of interdependencies among preferences in best matching, consider the following example (Fig. 7.13). John has just arrived at a social gathering and is going to choose a table (A, B, or C) at which to sit. Besides his initial preference for each individual table (e.g., location; food; lighting), his choice may be influenced by his interpersonal relations/emotions towards the people who are already sitting at each table. Attendance of John to each table may drastically

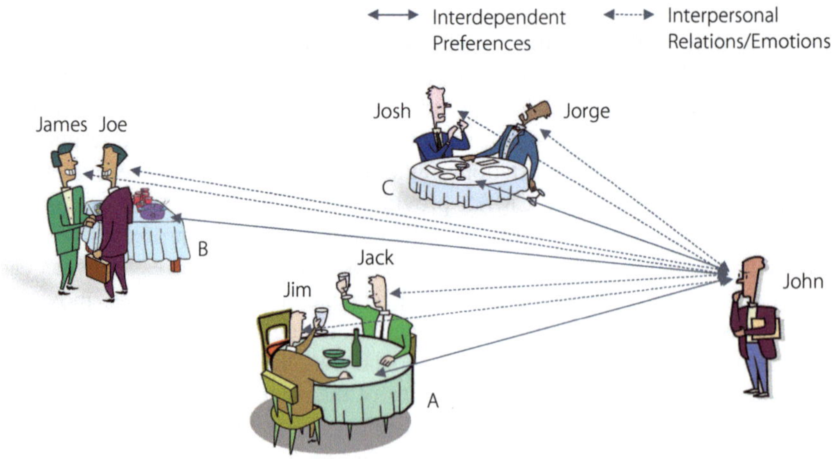

Fig. 7.13 The social example—the initial preferences of John for the tables (A, B, C) is influenced by his interpersonal relations/emotions with regard to the other guests already sitting at those tables. (after Moghaddam and Nof 2015d)

change the topology of the entire party. For instance, if John chooses Table C and Josh (who is already sitting at that table) is extremely *envious* of him, Josh may leave Table C right away. Similarly, if Joe is *altruistic* about John, he may leave Table B and join Table C. This is a natural phenomenon—we have different perceptions and attitudes about different people. The preferences, as described earlier, are merely an abstraction of various matching criteria, and thus, such interdependencies may have certain implications in different application domains. Some practical examples are:

- *Enterprise collaboration*. The *profitability* of a particular coalition (alliance) for an enterprise may be influenced (increased/decreased) by the members (i.e., other enterprises) of that coalition.
- *Wireless sensor networks*. The choices of an individual sensor for different clusters in terms of *energy consumption* may be influenced by the type, number, and energy level of the sensors already included in each cluster.
- *Swarm robotics*. The *efficiency* of an individual robot may be influenced by its assignment to different teams, depending on the type, number, and functionality of the robots in each team.
- *Scheduling*. The optimal allocation of a task to a machine in terms of *makespan* or *cost* may be influenced by the processing and/or setup requirements of the tasks that are already in process. or in the queue of each machine.
- *Storage assignment*. The best storage location for a particular product in terms of *total movement time* of material handling devices may be influenced by its *affinity* with the already allocated products.

Interdependent preferences (IP) considerably influence best matching decisions and lead to non-optimal or even paradoxical decisions, if disregarded. The notion of IP has been extensively investigated in utility theory, as an indication of the dependencies of the individuals' preferences on the *consumption* or *well-being* of other individuals in their neighborhood.[16] IP sometimes leads to *altruistic* or *envious* behaviors, where, instead of considering their own absolute payoffs, individuals tend to evaluate their payoffs *relative to* those of others. Examples of such behaviors include success/failure of collaborative marketing and sales teams; conflicts, job satisfaction, effectiveness, and turnover of interactive nursing units; efficiency and rate of errors/miscommunication in surgical units and operating rooms; performance, throughput, and cost of construction projects—all influenced by certain mutual interactions among individuals. This example defines, formulates, and analyzes an extension of the generalized matching with IP, where the elements of the same set influence each other's preferences for the elements of the other set (s), if matched to the same element.

The many-to-one best matching problem with IP (henceforth, BMP-IP) is presented here, where sets I and J are matched, considering interdependencies among

[16]Interested readers are referred to Tomes (1986), Koçkesen and Ok (2000), Bell (2002), Sobel (2005), Cabrales and Calvó-Armengol (2008), Lee et al. (2011), Morita and Burns (2014).

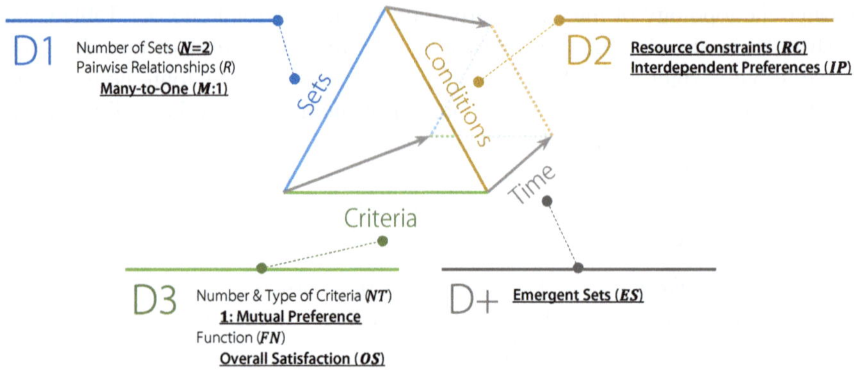

Fig. 7.14 Classification of the BMP-IP according to the PRISM taxonomy of best matching

the preferences of the elements of set I. The BMP-IP is a capacitated clustering problem, where each element of set J (i.e., the clusters) can be matched to a limited number of elements from set I. The clustering is performed with respect to the mutual preferences of I's and J's, the capacity limits of J's, and the influences of I's on each other's preferences. According to the PRISM taxonomy of best matching, the BMP-IP can be characterized as shown in Fig. 7.14. In this example, the BMP-IP is first investigated from the cluster formation perspective (i.e., without considering D+). The outcome of the capacitated many-to-one matching is a set of clusters, each corresponding to a specific element of set J. The problem is formulated as a binary program, with IP defined as a linear combination of the binary best matching variables. Due to the computational complexity of the BMP-IP, a genetic algorithm (GA) is developed next. Following cluster formation, the continuous review method presented in Chap. 6 (see Sect. 6.2.2) can be applied for optimal association/dissociation of individuals in real-time. The mechanism performs administration and control of the dynamic changes in the structure of the capacitated clusters (i.e., considering D+). These behaviors are also formalized by a recent principle of the collaborative control theory, the principle of *association/dissociation*. An evolutionary algorithm (EA) is presented for real-time optimization of the evolution of networked clusters based on the continuous review method presented in Chap. 6. Experiments show the impact of IP on the formation and evolution of capacitated clusters, the sensitivity of best matching decisions to the intensity of interdependencies among preferences.

7.4.1 Optimal Clustering: Genetic Algorithm (GA)

The mathematical formulation of the BMP-IP was presented earlier in Chap. 3 (see Model (M3.9)) as a quadratic assignment program (QAP). The QAP is NP-hard (Sahni and Gonzalez 1976), and so is the BMP-IP. Therefore, a GA is presented here, as a powerful population-based metaheuristic for solving the QAP[17]. The notion of GA is based on evolution through *natural selection*, which is also the foundation for the EA described later. A greedy heuristic is developed for generation of the initial population, and a reproduction scheme based on the *path relinking method* is introduced next for reproduction through crossover and mutation.

Encoding

The efficiency of a GA, in general, and its reproduction functions, in particular, highly depend on the encoding scheme deployed. For the BMP-IP, a natural encoding scheme is applied: an array of length $|I|$, where each specific gene on chromosome $\mathbf{c}$ corresponds to a specific element of set I, and the allele of each gene takes a value in J. Note that if the overall capacity of set J is less than the size of set I, some elements of set I may remain unmatched. Therefore, in the given encoding scheme, the alleles of the unmatched genes (if any) take zero values. Due to the limitations on the number and capacity of the elements of set J, a generated chromosome may be infeasible. Therefore, a greedy heuristic is developed for initialization of the algorithm, as described next.

Initialization

To ensure the diversity, fitness, and feasibility the initial population, a greedy heuristic is presented, which generates each individual chromosome through the following steps:

1. Generate a random permutation of the elements of I and store them in R, in that order.
2. Set $\hat{P}_{ij} \leftarrow P_{ij}, \forall i,j$.
3. Pick the first element of R, denote it by k, and assign it to $l \in J$ such that $\hat{P}_{kl} = \max_j\{\hat{P}_{kj}\}$.
4. Set $M_l \leftarrow M_l - 1$, and $R \leftarrow R \backslash \{k\}$.
5. Set $\mathbf{c}(k) = l$, $\chi_{kl} = 1$, and update all $\hat{P}_{ij}$, $\forall i,j$, using Eq. (3.12) in Chap. 3.
6. If $M_l > 0$, go to the next step. Otherwise, set $J \leftarrow J \backslash \{l\}$, and then go to the next step.
7. If

   (a) $R = \emptyset$, stop.
   (b) $R \neq \emptyset$ and $J = \emptyset$, set $\mathbf{c}(i) = 0, \forall i \in R$, and stop.
   (c) $R \neq \emptyset$ and $J \neq \emptyset$, return to Step 3.

The diversity of the initial solution is guaranteed by random ordering of the elements of set I while generating each chromosome (Step 1). The generated chromosomes are expected to have acceptable fitness values, because each individual is matched to its most-preferred "available" match (Step 3). The generation

[17]See Chap. 5 for detailed classification of metaheuristics and an introduction to GA.

also avoids capacity violation (i.e.,) by eliminating fully-loaded J's from the list of potential choices for the next I's (Steps 6 and 7-b).

Fitness Evaluation

The fitness value of each chromosome **c**, i.e., the overall satisfaction degree of mutual preferences/IP, is calculated using the objective function of Model (M 3.9) in Chap. 3, where $\chi_{ij} = 1$, if $\mathbf{c}(i) = j$, and $\chi_{ij} = 0$, otherwise.

Reproduction

After generation of the first population and evaluation of their fitness values, the next step is to produce the next generation from the current population of chromosomes. The "parent" chromosomes must be selected to produce offsprings. According to the theory of natural selection, reproduction must (usually) result in fitter individuals in the next generations (see Chapter 5). In this example, in addition to the fitness of the generated "offsprings", their feasibilities must also be considered. That is, infeasible solutions must be avoided during the reproduction procedure in order to uphold the performance of the algorithm. Thus, a reproduction scheme based on the path relinking method (Glover 1994) is applied as described next.[18]

- *Path crossover*. The crossover operator must probabilistically explore new solutions and exploit good traits of previous solutions. Hence, a modified version of the path crossover method with insert transformation (Ahujaa et al. 2000) is applied, which besides diversity and fitness, guarantees the feasibility of the offsprings (Fig. 7.15):

  1. Randomly select two parent chromosomes $\mathbf{c}_1$ and $\mathbf{c}_2$, and fix the genes with similar alleles (for exploitation of the good traits of parents). Randomly select a gene and set $Z = 0$.
  2. If the current gene is fixed, set $Z \leftarrow Z + 1$ and go to Step 3. Otherwise, perform *insert transformation*:

     (a) In parent $\mathbf{c}_1$, randomly select an unfixed gene (other than the current gene), which has the same allele as the one in the current gene of $\mathbf{c}_2$. If there is no such gene in $\mathbf{c}_1$, go to Step (c).
     (b) Insert the allele of the selected gene at the current gene of $\mathbf{c}_1$, and shift the alleles of other unfixed genes to the right.
     (c) Apply (a) and (b) to $\mathbf{c}_2$, and then go to Step (c).
     (d) Place the original parents and the generated offsprings into chromosome pool *CP*.

  3. If $Z = |I|$, go to Step 4. Otherwise, move to the next gene, from left to right in a cyclic fashion, and return to Step 2.

[18]See Chap. 5 for more detailed description of the reproduction operators of GA.

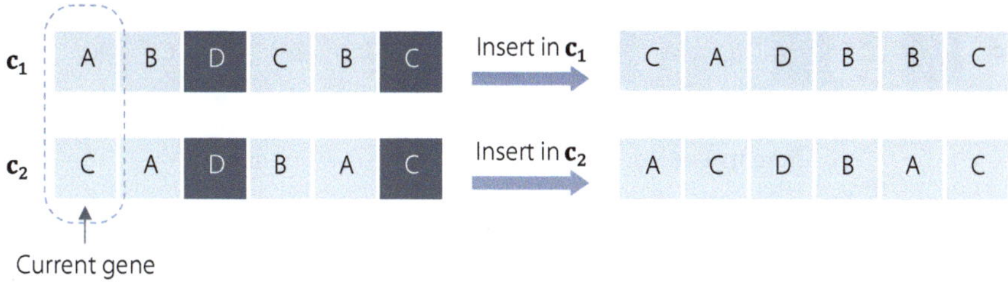

Fig. 7.15 Example of path crossover operation on two parent chromosomes $\mathbf{c}_1$ and $\mathbf{c}_2$ (*Dark shade* fixed genes)

4. Eliminate duplicate chromosomes from *CP* (if any), sort the rest according to their fitness values, and select a predefined number of the fittest chromosomes considering the population size and the crossover rate.

- *Path mutation.* The above path crossover mechanism does not make any changes in the consumed capacity of J's, because the number of I's matched to each $j \in J$ is the same in both parents and offsprings. This may eliminate some potential solutions. Hence, a *path mutation* mechanism is applied, which, similar to the crossover operator, produces offsprings through moving along a path and making spontaneous changes in the genes in sequence. The path mutation operator is designed in a way that the capacity limits of J's are not violated, while near-optimal offsprings are obtained. (If there is no idle capacity in any $j \in J$, the mutation operator is skipped.) The mutation procedure is as follows (Fig. 7.16):

1. Randomly select a parent chromosome $\mathbf{c}$ and a gene $k \in I$, set $Z = 0$, and generate a list of J's with free capacity, i.e., $L = \{j : M_j > |\{i : \mathbf{c}(i) = j, i \in I\}|, j \in J\}$.
2. Replace the allele of the current gene of $\mathbf{c}$ (i.e., k) with $l \in L$, where $\hat{P}_{kl} = \max_j\{\hat{P}_{kj}\}$.
3. Place the parent and the offspring into chromosome pool *CP*, and set $Z \leftarrow Z + 1$.
4. If $Z < I$, move to the next gene from left to right in a cyclic fashion, and return to Step 2. Otherwise, proceed to the next step.
5. Eliminate duplicate chromosomes from *CP* (if any), sort the rest according to their fitness values, and select a predefined number of the fittest chromosomes considering the population size and the mutation rate.

7.4.2 D+: Association/Dissociation

After each change takes place, the modified chromosome duplicates and reproduces itself using the GA operators to generate new offsprings and eventually a population

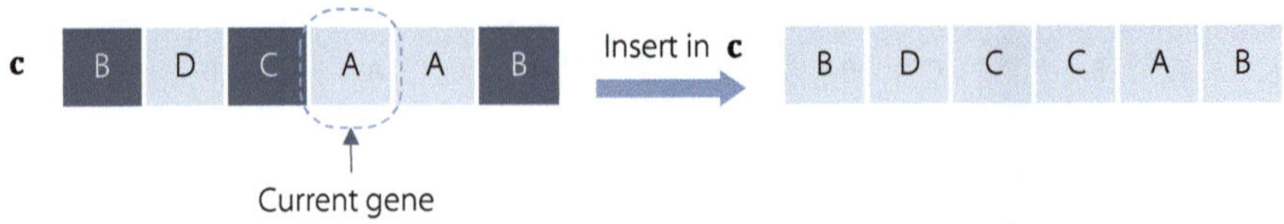

Fig. 7.16 Example of path mutation operation on parent chromosomes **c** (*Dark shade* $j \in L$)

of chromosomes.[19] The duplication and reproduction mechanism is analogous to the *mitosis* process in cell division and duplication (see Fig. 6.6 in Chap. 6). The modified chromosome first duplicates itself through mutation. The two chromosomes then generate two new chromosomes through crossover. The four chromosomes then generate four new chromosomes, and the procedure continues until the number of generated chromosomes reaches a predefined population size. Without loss of generality, the evolution from the modified chromosome toward the optimal one is based on the path crossover and path mutation mechanisms and the GA presented in the previous section. The EA is therefore composed of the following steps:

1. Define $C = \{\mathbf{c}^{\mathrm{m}}\}$ as the set of chromosomes in the first population, PS as the population size, and $L = \{j : M_j > |\{i : \mathbf{c}^{\mathrm{m}}(i) = j, i \in I\}|, j \in J\}$.
2. If $|L| > 0$, proceed to the next step. Otherwise, go to Step 5.
3. Perform path mutation on each chromosome $\mathbf{c}$ in C until an offspring chromosome $\mathbf{c}' \neq \mathbf{c}$ is generated. Set $C \leftarrow C \cup \{\mathbf{c}'\}$. If $|C| = PS$, go to Step 7. Otherwise, proceed to the next step.
4. Randomly pair the chromosomes in C. Perform path crossover on each pair $\mathbf{c}_1$ and $\mathbf{c}_2$ until two offsprings $\mathbf{c}_1' \neq \mathbf{c}_2' \neq \mathbf{c}_1 \neq \mathbf{c}_2$ are generated. Set $C \leftarrow C \cup \{\mathbf{c}_1', \mathbf{c}_2'\}$. If $|C| = PS$, go to Step 7. Otherwise, return to Step 3.
5. Perform swap mutation on $\mathbf{c}^{\mathrm{m}}$ to generate another chromosome. Set $\mathbf{c}' = \mathbf{c}^{\mathrm{m}}$, randomly select $i, k \in I$, $i \neq k$, and then $\mathbf{c}'(i) \leftarrow \mathbf{c}^{\mathrm{m}}(k)$ and $\mathbf{c}'(k) \leftarrow \mathbf{c}^{\mathrm{m}}(i)$. Set $C \leftarrow C \cup \{\mathbf{c}'\}$.
6. Randomly pair the chromosomes in C. Perform path crossover on each pair $\mathbf{c}_1$ and $\mathbf{c}_2$ until two offsprings $\mathbf{c}_1' \neq \mathbf{c}_2' \neq \mathbf{c}_1 \neq \mathbf{c}_2$ are generated. Set $C \leftarrow C \cup \{\mathbf{c}_1', \mathbf{c}_2'\}$. If $|C| = PS$, go to Step 7. Otherwise, return to Step 5.
7. Take C as the initial population, and run GA until the optimal configuration is obtained.

Steps 3–6 of the algorithm generate the first population of chromosomes. Steps 3 and 4 deals with the situations where at least one element of set J has extra capacity, and thus apply path mutation and path crossover. Steps 5 and 6, on the other hand, handle the situations where no element of set J has extra capacity (path mutation is not applicable). Hence, the modified chromosome first generates an offspring through a single swap mutation (see Fig. 5.6 in Chap. 5), and then the rest of the initial

[19]See the continuous review method of RTO in Chap. 6.

population is generated through path crossover. After generating the first population, the GA can be applied to find the optimal configuration of the evolved system.

7.4.3 Measured Impact

Experiments have proven the significant impacts of IP on the quality and accuracy of matching, in both formation of capacitated networks of clusters, and their evolution.[20] It is shown that relative to the traditional best matching with no IP, the BMP-IP results in better fitness values in cases where the interdependencies between preferences are significant (Fig. 7.17). This gap roots in the intensity of the influences of the elements of set I on each other's preferences; i.e., the α values. Experiments also show that this gap is a function of α, which implies that higher mutual influences of individuals on each other's preferences, either positive or negative, lead to more dramatic changes in the optimal network formation (Fig. 7.18). Moreover, as discussed in Chap. 6, IP plays a substantial role in optimal evolution (i.e., real-time optimization; D+) of the capacitated networks of clusters, and disregarding its roles may lead to "erroneous evolution" of the system (see Fig. 6.9 in Chap. 6).

7.4.4 Discussion

The BMP-IP models and analyzes the mutual influences of individuals on each other's preferences in generalized matching processes. It was shown through experiments that IP dramatically influences the generalized best matching decisions —the identified gap between the fitness values of best matching decisions with and without IP ranges from 5 to 25 %. This shows the substantial role that the notion of IP and the BMP-IP methodology play in improving the performance of any networked system with analogous features and structures—from homogenous teams of humans (e.g., pilots; soldiers; students; workers; technicians; doctors; interns; roommates) to heterogeneous teams of humans and/or machines, with applications in production, manufacturing, supply, logistics, healthcare, and transportation. The experiments indicate that the BMP-IP, besides all of its known inherent complexities and dynamics, is sensitive to the mutual *influences* between each pair of individuals. Although this phenomenon is more common in social networks compared to other types of systems, however, similar behaviors are expected from teams of artificial entities (e.g., sensors, robots, enterprises), based on factors analogous to emotion in humans (e.g., operational compatibility; task requirements). Product affinity (Li et al. 2016), for instance is a practical example of IP in storage allocation (see the examples

[20] The results presented here are the summary of the findings reported by Moghaddam and Nof (2015c, d).

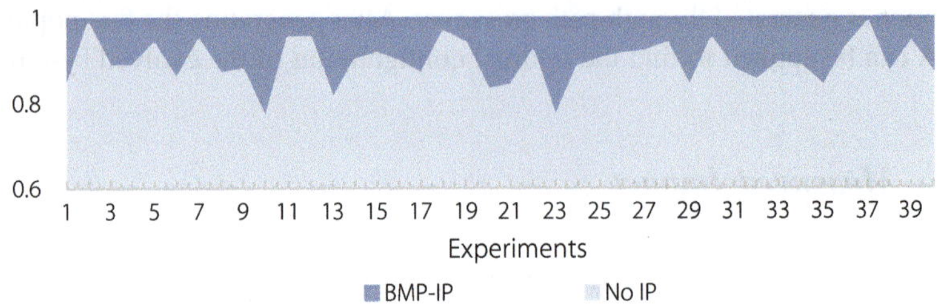

Fig. 7.17 Relative gap between the BMP-IP and the classic matching without considering IP in 40 independent experiments

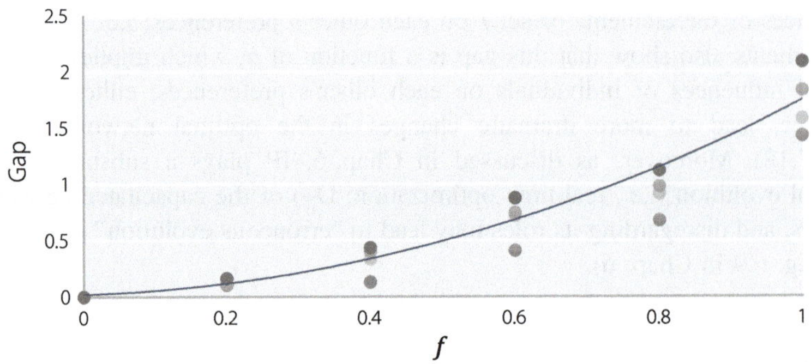

Fig. 7.18 Sensitivity of the gap between the normalized results of the BMP-IP and the classic matching without considering IP in different experiments. ($\alpha = [-f, f]$)

of storage assignment in Chap. 5), where products with higher affinity (i.e., the products that are ordered together more frequently) are located in the same/closest possible aisles of the warehouse so that the material handling equipment take shorter routes placing or picking up the products.

7.5 E_4: Collaborative Service Enterprises ($1{:}M{:}M/RC, RS/+-, WS$)

This case extends the example of location-allocation decisions in CNO (Collaborative Networked Organizations) presented in Chap. 2 (see Sect. 2.2.4). The purpose is to apply best matching models and algorithms for effective design and administration of CNO through optimal allocation of *tasks* to distributed *resources,* as well as optimal location of resources in networked *organizations* (see Fig. 2.9 in Chap. 2). As discussed in Chap. 2, a CNO, besides its physical dimension, also represents a *virtual* dimension; a distributed network of independent organizations and resources that collaborate to accomplish a set of individual and common goals. This virtual dimension enables collaboration between physical networked organizations—from minimal file exchange to direct access to resources such as computers, machines, software, and databases. That is, while some tasks require physical proximity to resources (e.g., sensing by distributed sensors; environmental monitoring or rescue by mobile robots; cross-docking in a distribution network), some other tasks can be processed remotely through information and communication channels (e.g., remote control of tele-robots; virtual manufacturing; remote modeling/simulation/problem-solving). In this example, the tasks of the former and the latter classes are respectively referred to as physical (p-)Tasks and electronic (e-)Tasks.

Definition 7.2 A CNO refers to a network of organizations that share information, resources, and responsibilities for accomplishing a set of tasks.[21]

In a CNO, due to the physical limitations associated with the p-Tasks as well as resources, collaboration is performed in both physical and virtual dimensions through two distinct but interrelated types of decisions (Fig. 7.1):

1. *Allocation of tasks* (virtual). In a CNO, each organization is responsible for processing a set of tasks using its "local" resources. Due to the dynamic variations in both demand and capacity, however, organizations may encounter capacity shortage/surplus over time (see Example E_1). Given the "e" and "p" classifications of tasks, organizations can collaborate by sharing their e-Tasks in case of insufficiency of local resources. Collaborative processing of e-Tasks balances the overall workload of the network, and minimizes idle resources, congestions, and delays.
2. *Location of resources* (physical). Unbalanced distribution of resources among organizations may lead to considerable inefficiency and diminish the positive impact of collaboration through e-Task sharing. For instance, some organizations may lack resources for processing their p-Tasks which require physical proximity to resources. Moreover, high rates of e-Task sharing among organizations, due to lack of local processing resources, leads to higher communication network load, congestion, delays, and cost, and increases the possibilities of errors and conflicts. Such challenges can be alleviated by enabling the organizations to dynamically share their "shareable" resources.

This example defines, formulates, and analyzes a collaborative location-allocation problem (CLAP) for designing CNO with respect to two (conflicting) objectives: (1) Maximizing task fulfillment rate; (2) Minimizing (unnecessary) collaboration rate. The objectives are achieved by synchronized best matching between the sets of tasks and resources, and the sets of resources and organizations. The CLAP is formulated as a bi-objective mixed-integer program. Due to the computational complexity of the CLAP (Owen and Daskin 1998), a tabu search algorithm is presented with a novel best matching heuristic for neighborhood search inspired by the natural *justice rule*. (Henceforth, the algorithm is referred to as TS-Jr.) The TS-Jr. algorithm enables the optimal (re)configuration of CNO given

[21]For detailed definition and classification of CNO, interested readers are encouraged to review Camarinha-Matos et al. (2009).

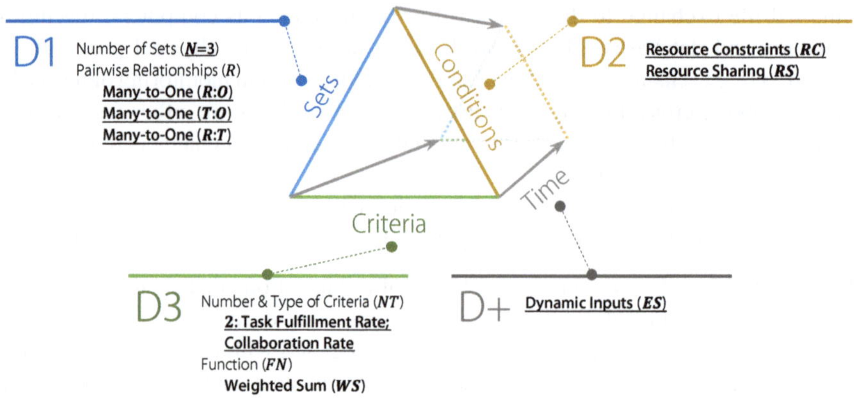

Fig. 7.19 Classification of the CLAP according to the PRISM taxonomy of best matching

the potential dynamic changes in the characteristics of the tasks and/or resources along with the topology of the network. Experiments show the significant impact of collaboration on the performance of CNO in terms of task fulfillment, utilization of resources, and network stability, which are in line with the emerging requirements of CNO for flexibility, scalability, and resilience. The PRISM taxonomy of best matching characterizes the CLAP as shown in Fig. 7.19.

7.5.1 Mathematical Formulation

Let $R_t \subseteq R$ denote the set of resources eligible for processing task $t \in T$, L_r denote the capacity limit of resource $r \in R$, C_{rt} denote the capacity required by task $t \in T$ if assigned to resource $r \in R_t$, T^p and T^e respectively denote the sets of p-Tasks and e-Tasks ($T = T^p \cup T^e$), and $T_o \subseteq T$ denote the set of tasks associated with organization $o \in O$. The CLAP can then be defined as a three-sided resource-constrained best matching problem between the elements of sets O, R, and T, with respect to two main criteria; task fulfillment rate and collaboration rate. A three-index formulation is used in the bi-objective mixed-integer program, where $\chi_{ort} = 1$, if task t is allocated to resource r, and resource r is located in organization o, and $\chi_{ort} = 0$, otherwise. The three-index formulation enhances the representation of the dependencies between the location and allocation decisions. The bi-objective program is formulated as follows[22]:

$$\text{Maximize} \quad Z_1 = \sum_{r \in R_t} \sum_{t \in T} \alpha_{rt}, \tag{7.30}$$

[22]This is an extension of Model (M 3.7), M: $1/RC, RS/+, OS$, presented in Chap. 3.

$$\text{Minimize} \quad Z_2 = \sum_{o \in O} \sum_{r \in R} \sum_{\substack{t \in T \\ t \notin T_o}} \chi_{ort} \alpha_{rt}, \tag{7.31}$$

Subject to

$$\sum_{t \in T} \chi_{ort} \alpha_{rt} \le L_r, \quad \forall o \in O, r \in R, \tag{7.32}$$

$$\sum_{r \in R_t} \left(\frac{\alpha_{rt}}{C_{rt}} \right) \le 1, \quad \forall t \in T, \tag{7.33}$$

$$\alpha_{rt} \le C_{rt} \sum_{o \in O} \chi_{ort}, \quad \forall r \in R, t \in T, \tag{7.34}$$

$$\chi_{ort} = 0, \quad \forall o \in O, r \in R, t \in T^p, t \notin T_o \quad or \quad \forall o \in O, r \notin R_t, t \in T, \tag{7.35}$$

$$\sum_{\substack{o' \in O \\ o' \neq o}} \sum_{t' \in T} \chi_{o'rt'} \le M(1 - \chi_{ort}), \quad \forall o \in O, r \in R, t \in T, \tag{7.36}$$

$$\alpha_{rt} \ge 0, \chi_{ort} \in \{0, 1\}, \forall o \in O, r \in R, t \in T. \tag{7.37}$$

In Objective Function (7.30), α_{rt} denotes the amounts of resources $r \in R_t$ consumed by task $t \in T$. Thus, the fulfillment rate (FR) of each individual task can be estimated using the following formula:

$$FR_t = \sum_{r \in R_t} \left(\frac{\alpha_{rt}}{C_{rt}} \right), \quad \forall t \in T, \tag{7.38}$$

Objective Function (7.31) minimizes the total amount of task sharing; i.e., the usage of external resources associated with other organizations for processing local task. This function must take zero value for all p-Tasks. (This requirement is addressed Constraints (7.35).) Given the value of this function, the collaboration rate (CR) can be estimated using the following formula:

$$CR = \frac{Z_2}{\sum_{r \in R} \sum_{t \in T} \alpha_{rt}}, \tag{7.39}$$

In this context, CR represents the ratio of the total collaboratively-processed/-consumed tasks/resources to the overall processing/consumption of tasks/resources. Constraints (7.32) ensure that the capacity limits of the resources are not violated. Constraints (7.33) ensure that the FR of each task is bounded to one. Constraints (7.34) implies that if task t is not assigned to resource r, the respective consumption value α_{rt} must be zero; otherwise, α_{rt} must be limited to C_{rt} (i.e., the capacity required by task t if processed by resource r). Constraints (7.35) ensure that (i) all p-Tasks are only processed by their respective organization, and (ii) the tasks (both "e" and "p" classes) are not assigned to ineligible resources. Constraints (7.36) satisfy the condition that each individual resource cannot be physically located in multiple organizations (M is a sufficiently large positive number). Constraints (7.37) ensure the feasibility of the decision variables.

7.5.2 *Optimization: Tabu Search*

The CLAP is a multidimensional generalized matching problem, and is NP-hard (Owen and Daskin 1998). Various exact, approximation, relaxation, and heuristic algorithms have been proposed in literature for solving multidimensional generalized matching problems.[23] An effective alternative amongst all those algorithms in terms of both solution quality and computational efficiency is tabu search.[24] As a memory-based neighborhood search metaheuristic with *prohibition* strategies, tabu search marks previously visited solutions as (temporary or permanent) forbidden or *tabu* points in order to prevent cycling and improve the efficiency of search mechanism. The algorithm starts with an initial solution S, and explores through its neighborhood while exploiting the knowledge of tabu points/areas, until a stopping criterion is reached. The moves are directed by various mechanisms, among which *shift*, *swap*, and *ejection chains* (Glover 1996) are the most common. For the multidimensional best matching problem, specifically, simple, double, and circular ejection chains are recommended as efficient mechanisms for movement in the neighborhood (Laguna et al. 1995). Besides the movement mechanism, however, a critical step in tabu search is to find the "best move" in each iteration. In the CLAP, due to the interdependencies between task sharing and best matching decisions, identification of the best move may not be possible via simple infeasibility or cost-improvement measures. Thus, a recent TS-Jr. algorithm is introduced here which is composed of two main stages for generating a feasible, high-quality initial solution S, and identifying the best next move to generate an updated solution S' (Moghaddam and Nof 2015e). The two stages of the TS-Jr. algorithm are described below.

Initialization: The Rock and Sand Heuristic

The performance of tabu search is significantly influenced by the feasibility and quality of the initial solution S. In the CLAP, feasibility implies satisfaction of Constraints (7.32)–(7.37). An encoding scheme is therefore applied which guarantees satisfaction of Constraints (7.34) to (7.37). A heuristic procedure inspired by

[23]See, for example, Lorena and Narciso (1996), Yagiura et al. (1999), De Farias et al. (2000), Lourenc and Serra (2002), Haddadi and Ouzia (2004), Yagiura et al. (2004), Cohen et al. (2006), Jeet and Kutanoglu (2007).

[24]See Chap. 5 for detailed classification of metaheuristics and an introduction to tabu search.

the *rock and sand analogy* in time management is also applied, which besides satisfaction of the rest of the constraints, improves the quality for the initial solution S. The idea of the rock and sand initialization heuristic is to allocate larger resources (≡rocks) first to the organizations with higher demand, and then allocate smaller resources (≡sands) to enhance the balance between resources and demand, and thus, minimize the CR. Allocation of resources to different tasks is then performed such that the FR is maximized, while preserving the resource limits. The rock and sand initialization heuristic is as follows:

0. Generate a $(|T|+1) \times |R|$ zero matrix $\mathbf{M}$.
1. Estimate the overall resource requirement of each organization:

$$ER_o = \sum_{t \in T_o} \sum_{r \in R_t} \left(\frac{C_{rt}}{|R_t|} \right), \quad \forall o \in O. \tag{7.40}$$

2. Locate the largest resource in the organization with maximum ER_o; i.e., $\mathbf{M}[1, \hat{r}] \leftarrow \hat{o}$, where $L_{\hat{r}} = \max_r\{L_r\}$ and $ER_{\hat{o}} = \max_o\{ER_o\}$. (Break ties arbitrarily.) Update $ER_{\hat{o}} \leftarrow ER_{\hat{o}} - L_{\hat{r}}$ and $R \leftarrow R \backslash \{\hat{r}\}$.
3. If $R = \emptyset$, proceed to Step 4. Otherwise, return to Step 2.
4. Given the location decisions, estimate the average resource requirement of each task:

$$AR_t = \sum_{r \in X_t} \left(\frac{C_{rt}}{|R_t|} \right), \quad \forall t \in T, \tag{7.41}$$

   where $X_t = \{r : r \in R_t; \mathbf{M}[1, r] = o; t \in T_o\}$, $\forall t \in T^p$, and $X_t = R_t$, $\forall t \in T^e$.
5. If $T = \emptyset$ or $R = \emptyset$, go to Step 7. Otherwise, identify task $\hat{t}$ where $AR_{\hat{t}} = \min_t\{AR_t\}$ and set $z_{\hat{t}} = 1$. (Break ties by prioritizing p-Tasks over e-Tasks; if the tasks are the same type, break ties arbitrarily.)
6. Identify resource $r \in X_{\hat{t}}$ where $C_{\hat{r}\hat{t}} = \max_r\{C_{rt}\}$. (Break ties arbitrarily.) Calculate $\alpha_{\hat{r}\hat{t}} = z_{\hat{t}} C_{\hat{r}\hat{t}}$:
   - If $\alpha_{\hat{r}\hat{t}} < L_{\hat{r}}$, set $\mathbf{M}[\hat{t}+1, \hat{r}] \leftarrow \mathbf{M}[\hat{t}+1, \hat{r}] + \alpha_{\hat{r}\hat{t}}$, $L_{\hat{r}} \leftarrow L_{\hat{r}} - \alpha_{\hat{r}\hat{t}}$, $T \leftarrow T \backslash \{\hat{t}\}$, and return to Step 5.
   - If $\alpha_{\hat{r}\hat{t}} > L_{\hat{r}}$, set $\mathbf{M}[\hat{t}+1, \hat{r}] \leftarrow \mathbf{M}[\hat{t}+1, \hat{r}] + L_{\hat{r}}$, $L_{\hat{r}} \leftarrow 0$, $R \leftarrow R \backslash \{\hat{r}\}$, $X_{\hat{t}} \leftarrow X_{\hat{t}} \backslash \{\hat{r}\}$, and $z_{\hat{t}} \leftarrow z_{\hat{t}} - (L_{\hat{r}}/C_{\hat{r}\hat{t}})$. If $X_{\hat{t}} \neq \emptyset$, redo Step 6; otherwise, return to Step 5.
7. Output matrix $\mathbf{M}$ as the initial solution S.

The rock and sand heuristic is composed of two phases for location (Steps 1–3) and allocation (Steps 4–6) decisions. The algorithm estimates the amount of resources required by each organization, and locates the largest resources, one at a time, in the organization with the highest demand (Steps 2 and 3). This procedure minimizes the CR by distributing the resources with respect to the distribution of demand. The idea behind the task allocation heuristic (i.e., Steps 4-6) is that tasks with lower average resource requirements, in general, have higher priority for allocation, and in case of ties, p-Tasks are prior to e-Tasks. The algorithm therefore prioritizes the tasks based on their average resource requirement, and then allocates them, one at a time, to the most efficient resource(s) available. The procedure continues until all tasks and/or resources are allocated. The fitness of matrix **M** can be calculated using the following formula:

$$F(\mathbf{M}) = w_{FR}\left(\frac{\sum_{t\in T} FR_t}{|T|}\right) + w_{CR}(1 - CR), \tag{7.42}$$

where w_{FR} and w_{CR} denote the weights of the FR and CR objectives, respectively. The values of FR_t and CR are respectively obtained using Eqs. (7.2) and (7.4), where $\chi_{ort} = 1$, if $\mathbf{M}[1, r] = o$ and $\mathbf{M}[t+1, r] > 0$, and $\chi_{ort} = 0$, otherwise, and $\alpha_{rt} = \mathbf{M}[t+1, r]$, $\forall o \in O, r \in R, t \in T$.

Neighborhood Search: The Natural Justice Rule

The theory of Social Darwinism, an extension of Charles Darwin's law of natural selection to sociology, stresses the existence of justice in nature by stating that stronger/weaker individuals or groups have higher/lower chances of survival, growth, and dominance in society—that *individuals gain what they deserve*. Oxford Dictionaries define *justice* as "just behavior or treatment" or "the quality of being fair and reasonable", which is reflected by the notion of the natural justice rule: an individual's "gain" is an indication of their "power". In the context of matching problems, an individual's power may be represented by their overall popularity, associated costs/benefits, resources, fitness, dimensional tolerance, *etc*. Therefore, following the natural justice rule, the *deviation* between the power and the gain of individuals can be viewed as an indication of the quality of a solution (Moghaddam and Nof 2015e). In tabu search, the moves in each iteration are applied in the neighborhood by making slight changes in the current solution S; e.g., changing the allocation of one or more tasks. Nevertheless, the challenge is to find the such that the algorithm converges faster. The neighborhood search heuristic introduced here addresses this issue, based the natural justice rule, by defining a generic *power-gain deviation function* as follows:

$$D_i = P_i - G_i, \tag{7.43}$$

where P_i and G_i denote the power and the gain of individual i, respectively. Thus, individuals with larger $|D_i|$ have higher likelihood of being selected for the next move in the neighborhood. For instance, if the actual processing time of a task on its assigned machine (i.e., gain) is considerably higher/lower than its average processing time on all machines (i.e., power), there is a high likelihood that this is not an optimal match, neither for the task nor for the machine, and thus, there is high chance that the task/machine is selected for the next move. The key point, however, is to properly define the power and gain values in different best matching instances. In the CLAP, the natural justice rule is applied to set O (organizations) regarding their matching to the elements of sets R (resources) and T (tasks). The power-gain deviation function for the elements of set O is therefore calculated using the following formula:

$$D_o = \underbrace{\sum_{t \in T_o} \sum_{r \in R_t} \left(\frac{C_{rt}}{|R_t|} \right)}_{P_o} - \underbrace{\sum_{r \in R} L_r \max_t \{\chi_{ort}\}}_{G_o}, \quad \forall o \in O, \tag{7.44}$$

D_o is a function of the deviation between the total resource requirements of the tasks attributed to organization o (i.e., P_o), and the total level of resources assigned to organization o (i.e., G_o). Minimization of $|D_o|$ in an iterative fashion can then automatically direct the neighborhood search towards minimization of the CR. In order to maximize the FR objective, the rock and sand heuristic is applied for reallocation of the tasks in each iteration. The neighborhood search heuristic is as follows:

0. Set matrix $\mathbf{M}_0 = \mathbf{M}$ as the initial solution S, $TL = \emptyset$ as the initial tabu list, $IT = 0$ as the counter for stopping the algorithm, and $\tau(o, r, t)$ as the tabu time of the triplet (o, r, t), $\forall o \in O, r \in R, t \in T$.
1. Randomly generate $\gamma \in [0, 1]$. If $\gamma < \gamma_0$, select organizations $\hat{o}$ and $\breve{o}$ where $|D_{\hat{o}}| = \max_o\{|D_o|\}$ and $|D_{\hat{o}} - D_{\breve{o}}| = \max_o\{|D_{\hat{o}} - D_o|\}$. Otherwise, randomly select organizations $\hat{o}, \breve{o} \in O$, $\hat{o} \neq \breve{o}$.
2. If $D_{\hat{o}} > D_{\breve{o}}$, identify resource $\hat{r}$ where $\mathbf{M}[1, \hat{r}] = \breve{o}$ and $|L_{\hat{r}} - |D_{\hat{o}}|| = \min_{r:\mathbf{M}[1,r]=\breve{o}}\{|L_r - |D_{\hat{o}}||\}$, and update $\mathbf{M}[1, \hat{r}] \leftarrow \hat{o}$. Otherwise, identify resource $\hat{r}$ where $\mathbf{M}[1, \hat{r}] = \hat{o}$ and $|L_{\hat{r}} - |D_{\breve{o}}|| = \min_{r:\mathbf{M}[1,r]=\hat{o}}$ $\{|L_r - |D_{\breve{o}}||\}$, and update $\mathbf{M}[1, \hat{r}] = \breve{o}$.
3. Follow Steps 4–6 of the rock and sand heuristic to reallocate all the tasks based on the updated location of the resources. For each task $\hat{t}$, remove resource $\hat{r}$ from the list of eligible resources, i.e., $X_{\hat{t}} \leftarrow X_{\hat{t}} \backslash \{\hat{r}\}$, if their combination belongs to the tabu list, i.e., $(o, \hat{r}, \hat{t}) \in TL$, where $\mathbf{M}[1, \hat{r}] = o$.
4. Output matrix $\mathbf{M}$ as the updated solution S'. If $0 < F(\mathbf{M}) - F^* < \delta$, $\delta > 0$, update $IT \leftarrow IT + 1$, where F^* denotes the best solution so far; otherwise, set $IT = 0$. If $IT > IT_{\max}$, stop and output $\mathbf{M}$ as the optimal solution; otherwise, go to Step 5.
5. Update all positive tabu times $\tau(o, r, t) \leftarrow \tau(o, r, t) - 1$, $\forall o \in O, r \in R, t \in T$ and $\tau(o, r, t) > 0$, and

$$\begin{aligned} \tau(o, r, t) = {} & \max\{\lceil \theta(F(\mathbf{M}) - F(\mathbf{M}_0)) \rceil, 0\}, \\ & \forall (o, r, t) \\ & \quad \in \{(o, r, t) : \mathbf{M}[t+1, r] = 0, \mathbf{M}_0[t+1, r] = 0, \mathbf{M}_0[1, r] = o\}, \end{aligned} \tag{7.45}$$

where $\theta > 0$ is a coefficient, and $\lceil \cdot \rceil$ denotes the ceiling function.
6. Update the tabu list $TL \leftarrow TL \cup \{(o, r, t)\}$, $\forall o \in O, r \in R, t \in T$ and $\tau(o, r, t) > 0$, set $\mathbf{M}_0 = \mathbf{M}$ as the updated solution S, and return to Step 1.

The heuristic *probabilistically* selects two organizations in two consecutive rounds, following the natural justice rule. In the first round, an organization with the largest deviation is selected, while in the second round, an organization with the most opposite situation of the first organization is selected. Parameter γ_0 enables exploitation of the knowledge provided by the natural justice rule as well as exploration of new solutions in each iteration. The selection is based on the ejection chains method (Laguna et al. 1995), where the organization with extra capacity shares one of its resources that best fits the resource requirement of the organization with capacity shortage (Step 2). Then, the tasks that are not on the tabu list are reallocated to the organizations and resources via the task allocation module of the rock and sand heuristic introduced earlier (Steps 3). The fitness of the updated solution is then calculated. If no improvement is made over the last $IT_{\max}$ iterations, the algorithm stops (Step 4). Otherwise, it updates the tabu times and the tabu list (Step 5), and resumes the neighborhood search (Step 6). Without loss of generality, the tabu times are considered proportional to the improvements made in the fitness values after each iteration. If there is no/negative improvement, the leaving variables do not enter the tabu list.

7.5.3 Measured Impact

Experiments show that the CLAP significantly improves the performance of CNO in terms of task fulfillment and collaboration efficiency.[25] In addition, the TS-Jr. algorithm is proven to be capable of making (near) optimal decisions on the location of resources, allocation of tasks to resources/organizations, and collaboration among organizations through resource sharing and task sharing. Results show that the CLAP significantly outperforms the equivalent non-collaborative scenario in terms of FR (Fig. 7.20), due to the possibility of collaborative processing of e-Tasks. This indicates the significant impact of collaboration on the service level and utilization of distributed resources, where less than 20% of the e-Tasks, on average, are processed locally at their respective organizations. This, in turn, increases the flexibility and stability of CNO in handling demand disruptions with the same level of resources (see Example E_1). Experiments also show that as the number of resources and tasks—and thus their variety—increases, the organizations become even more capable of fulfilling their tasks (compared to the non-collaborative scenario) due to the access to a larger and more diverse pool of resources and higher likelihood of finding their best match. This phenomenon can be explained, more technically, by the collaborative fault tolerance principle of the

[25]The results presented here are the summary of the findings reported by Moghaddam and Nof (2015c).

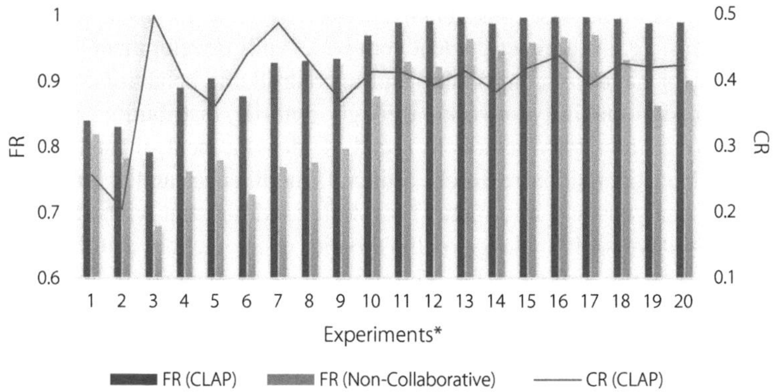

Fig. 7.20 Experiments on the FR and CR values—CLAP versus non-collaborative scenario (FR only). (*: The experiments are sorted in ascending order of problem size; see Moghaddam and Nof 2015e.)

CCT (Nof 2007), where larger networks of agents/resources are proven to result in more resilient network—and thus higher FR—through *teaming*.

7.6 Summary

Advances in electronics and communication technologies have gradually increased the need for scalability, interoperability, integrability, and resilience of systems, and enabled concurrency in distributed operations. The escalated rates of interactions in such complex networks, however, have brought up a new cohort of challenges in terms of flexibility and scalability of systems, along with optimality and timeliness of decisions. In this context, the hypothesis is that if every single element of a distributed system is matched to its best peers at the right time (e.g., suppliers, retailers, and customers in enterprise networks; jobs, machines, and sensors in factories; vehicles and routes in transportation systems; robots, humans, and orders in warehouses), the outcome will be a better system that addresses such emerging challenges. The complex, dynamic, and uncertain nature of modern manufacturing and service systems, however, has made such best matching processes very difficult to accomplish; in terms of both distribution of decisional capabilities (e.g., hierarchy/heterarchy) and nature of interactions among distributed individuals (e.g., competitive/cooperative).

Researchers and practitioners must investigate the key success factors in interconnected and advanced manufacturing, supply, and service networks, develop solutions for real-time optimization and control of interactions, and enhance the individuals' intelligence for timely decision-making regarding *what to do*, *with whom to interact*, *how* and *when*. Realization of this goal indeed requires (1) comprehensive characterization and taxonomy of distributed manufacturing and service systems, their requirements, constraints, and objectives; (2) assessment of the nature of interactions (e.g., competitive/cooperative), and development of proactive mechanisms for handling controllable and uncontrollable behaviors; (3) design of effective decision-making networks through optimal distribution of decisional capabilities (e.g., hierarchy/heterarchy); and (4) development of models and algorithms based on operations research, artificial intelligence, and information technology (i.e., analytics and informatics) for design, administration, and feedback.

The current level of knowledge on the theories of interaction, distributed control, and matching is not sufficient for addressing such challenges in modern manufacturing and service systems, and that motivates research on interaction engineering through best matching. The PRISM model formalizes such processes with respect to 3+1 dimensions, and provides a comprehensive and standardized framework for identification, specification, and design, in various domains such as resource sharing and supply (Example E_1), design and manufacturing (Example E_2), social organizations (Example E_3), and service quality (Example E_4), all modeled as networks. This chapter highlights the significant impact of best matching on the competitive performance of distributed systems through the four extended examples. The lessons learned and the research directions identified out of these four extensive cases are summarized below.

1. *Variability and uncertainty*. The uncertainties associated with costs of matching/sharing, information exchange, and possibility of acceptance/rejection of collaboration proposals during negotiations are practical scenarios that must be systematically addressed in future research.
2. *Steady state matching*. Optimal cluster design (e.g., for supplier-customer/tool-workstation pairs) in the steady state, and inter- and intra-collaboration decisions and protocols are other directions to be addressed in future research.
3. *Real-time decisions*. The four cases present, a "macro" view of matching processes. Implementation of such algorithms and protocols in real time, however, requires certain information exchange and negotiation procedures between different individuals. This issue must be addressed in future work.
4. *Communication and informatics*. Future research also must address the design of middleware architecture and the supportive components, decision support systems, modeling tools, and database systems associated with the presented best matching processes.
5. *Conflict and error detection and prevention*. Future research must address the issues related to the detection, resolution, and prevention of potential errors and conflicts in matching; e.g., delayed response/sharing; and conflict of interest or motivation for collaboration.
6. *Group- versus self-orientation*. It is assumed in all examples that all individuals have "incentives" for collaboration. That is, all elements are *group-oriented*, seeking a set of common objectives that are necessarily in line with their local objectives. In some cases, however, the individuals may be *self-oriented* and their individual goals may contradict the common goal of the network. The

network is then prone to instability, and the developed methodologies may require modifications in order to be applicable.

7. *Variable preferences and perceptions*. The preferences of individuals (e.g., cost; time) are modeled as functions of the mutual influences of individuals on each other's preferences, where each pair of individuals is assumed to have fixed *perceptions* about each other (see E_3). This assumption, however, may not be true in situations where two individuals may change their perceptions about each other after a period, or dynamically over time. This issue increases the complexity of the match and requires modifications in the definitions, formulations, and methodology.
8. *Multidimensional matching*. The basic instance of generalized matching was considered for studying the BMP-IP for the sake of simplicity in definitions and formulations. In a similar manner, the methodology can be extended to more comprehensive and realistic instances of the BMP-IP.
9. *Social networks and emotion*. An important trait that differentiates networks of humans (social networks, organizations) from other types of networks is *emotion*. The IP was formulated as a linear (increasing, altruism; decreasing, envy) function of the mutual influences of individuals on each other's preferences. In social networks, however, such influences may not be easily quantifiable and may have nonlinear relationship with IP, due to the complexity and dynamicity of humans' emotion in their mutual communications and interactions.
10. *Emergence of "coopetition"*. Despite the benefits of collaboration, competition is an inevitable behavioral pattern in almost every system, from microorganisms to multinational corporations. Altruistic and selfish behaviors always go side by side and give each other meaning. The notion of "coopetition" is then coined in game theory to represent situations where competitors prefer to both compete and collaborate with each other. In the context of matching, this is an important topic to address in the future, where the definition of "the best match" is dynamic and depends on the mutual interaction and local benefits of individuals.

Appendix 1: Notation[26]

E_1: *Collaborative Supply Networks*

Indices

i Customer ($i \in I$)
j,j' Supplier ($j,j' \in J$)
t Period ($t \in T$)

[26]**Note**: The notations are listed separately for each example, because some are repeated in different examples with different definitions.

Parameters

α_i Quantity of order from customer i
b_{jt} Backordering cost of supplier j in period t ($/unit)
$B_{j,0}$ Initial backorder level of supplier j at the beginning of planning horizon (units)
D_{it} Demand of customer i in period t (units)
δ_i Due date of order from customer i
$f_{jj't}$ Fixed cost of collaboration between suppliers j and j' in period t ($)
h_{jt} Inventory holding cost of supplier j in period t ($/unit)
$H_{j,0}$ Initial inventory level of supplier j at the beginning of planning horizon (units)
K_{jt} Capacity limit of supplier j in period t (units)
o_i Order from customer i
p_{jt} Production cost of supplier j in period t ($/unit)
π Priority
τ_i Processing time of order from customer i
$v_{jj't}$ Cost of transshipping products suppliers j and j' in period t ($/unit)
M A sufficiently large positive number

Variables

B_{jt} Backorder level of supplier j at the end of period t (units)
H_{jt} Inventory level of supplier j at the end of period t (units)
Q_{jt} Production level of supplier j in period t (units)
$S_{jj't}$ Transshipment level between suppliers j and j' in period t (units)
χ_{ijt} 1, if customer i is matched to supplier j in period t; 0, otherwise
$\lambda_{jj't}$ 1, if capacity sharing proposal of supplier j is matched to demand sharing proposal of supplier j' in period t; 0, otherwise

E_2: Collaborative Assembly Lines

Indices

i, i' Workstation ($i, i' \in I$)
j, j' Task ($j, j' \in J$)

Parameters

$\alpha_i(t)$ Progress rate of workstation i at time t
$\bar{C}$ Cycle time upper bound
C^B Cycle time of balanceable line
D Demand
δ Deviation of cycle time from C^B
$e_{ii'}$ CE for tool sharing from i to i'
$f_{ii'}$ Fixed tool sharing cost from i to i'
Pd_j Set of immediate predecessors of task i

$O_k(t)$ Set of non-target workstations with higher workload than k's targets at time t
ρ_{jt} Processing time of task j at time/period t
A_t Available production time in period t
$\overline{W}$ Upper bound for the number of workstations
$\underline{W}$ Lower bound for the number of workstations
$w_i(t)$ Workload of workstation i at time t

Variables

C Cycle time
$S_{ii't}$ Tool sharing between workstations i and i' in period t
W Number of workstations
χ_{jit} 1, if task i is matched to workstation i in period t; 0, otherwise
$\psi_{ii'}$ 1, if workstation i shares tools with workstation i' in period t; 0, otherwise
Z_1 Objective 1: Number of workstations
Z_2 Objective 2: Cycle time
Z_3 Objective 3: Total collaboration cost

E_3: Clustering with Interdependent Preferences
Indices

i, k Set I
j, l Set J

Parameters

$\alpha_{ii'}$ Influence of i' on the preferences of i
c Chromosome (solution set)
$\mathbf{c}^{\mathrm{m}}$ Modified chromosome
M_j Capacity of j
$p_{i \to j}$ Preference of i for j
P_{ij} Mutual preference of i and j
$\hat{P}_{ij}$ IP of i and j

Variables

χ_{ij} 1, if i and j are matched; 0, otherwise

E_4: Collaborative Service Enterprises
Indices

o Organizations ($o \in O$)
R_t Set of eligible resources for processing task t
r Resources ($r \in R$)
t Tasks ($t \in T$)
T_o Set of tasks associated with organization o
TL Tabu list

Parameters

AR_t	Average resource requirement of task t
C_{rt}	Capacity required by task t if processed by resource r
D_o	Deviation between power and gain of organization o
δ	Minimum deviation between the current and best fitness values
ER_o	Estimated (average) overall resource requirement of organization o
G_o	Gain of organization o (total resources assigned to o)
$IT; IT_{\max}$	Number of iterations; Stopping criterion
L_r	Capacity limit of resource r
γ_0	Exploration/exploitation probability in neighborhood search
$\mathbf{M}$	Encoding matrix
P_o	Power of organization o (average requirements of tasks from o)
θ	Tabu time function coefficient
$\tau(o, r, t)$	Tabu time of triplet (o, r, t)
w_{FR}	Weight of the FR objective
w_{CR}	Weight of the CR objective

Variables

α_{rt}	Amounts of resources r consumed by task t
CR	Collaboration rate
F	Fitness value
FR_t	Fulfillment rate of task t
χ_{ort}	1, if task t, resource r, and organization o are matched; 0, otherwise

Problems

7.1. Beyond the resource sharing cases discussed in this chapter, suggest other resource sharing applications and opportunities that can benefit from best matching.

7.2. What measurements would you apply to compare between resource sharing with best matching, and without it?

7.3. Distinguish between demand-capacity sharing by best matching for:

(a) Physical product supply.
(b) Digital product supply.
(c) Service delivery (e.g., transportation; food delivery; *etc.*).

7.4. Explain how you can benefit from the techniques of E_1 in the following cases of shared economy:

(a) Order delivery by shared drones.
(b) Repair of failed street lamps by shared maintenance-robots.
(c) Shared taxi service.

7.5. When would simulation be preferred to optimization heuristics in the solution of the four extended example problems in this chapter?

7.6. For the TAP described in Sect. 7.2.2, explain:

(a) What is the objective of applying a time-out?
(b) What are the disadvantages of applying time-out?
(c) Formulate an equation to quantify and compare the benefit and the loss due to the application of time-out.
(d) Can you suggest a protocol that can work effectively without a time-out?

7.7. For the TAP described in Sect. 7.2.2, develop three other protocol logic examples, as follows, and compare between them:

(a) Fixed priorities.
(b) Shortest Processing Time (SPT) first.
(c) An evolutionary (learning) logic.

7.8. Explain proposition 7.1 in your own words, and illustrate it with a numerical example.
7.9. Consider the negotiation function in Fig. 7.3, describe several alternative negotiation procedures for it, and define evaluation metrics for each of them.
7.10. What is the influence of the negotiation logic on measured impacts (as described in 7.2.3)?
7.11. In the workflow of the decentralized collaboration procedure shown in Fig. 7.4, explain the role of the best matching algorithms explained in Chaps. 5 and 6.
7.12. Repeat Problem 7.2 for the cases described in E_2, E_3, E_4.
7.13. Explain how best matching specifically impacts each of the four extended examples in this chapter. In your answer, focus on:

(a) The metrics of these impacts;
(b) How can these metrics be measured?

7.14. (a) Develop your own extended example, E_5, which is different from the four described in this chapter. (b) Use a table to compare the main modeling features of the five extended examples. (Hint: Use the PRISM taxonomy in your comparison.)

References

Ahujaa, R. K., Orlinb, J. B., & Tiwaric, A. (2000). A greedy genetic algorithm for the quadratic assignment problem. *Computers & Operations Research, 27*, 917–934.

Barbati, M., Bruno, G., & Genovese, A. (2012). Applications of agent-based models for optimization problems: A literature review. *Expert Systems with Applications, 39*, 6020–6028.

Becker, C., & Scholl, A. (2006). A survey on problems and methods in generalized assembly line balancing. *European Journal of Operational Research, 168*, 694–715.

Bell, A. M. (2002). Locally interdependent preferences in a general equilibrium environment. *Journal of Economic Behavior & Organization, 47*, 309–333.

Boysen, N., Fliedner, M., & Scholl, A. (2007). A classification of assembly line balancing problems. *European Journal of Operational Research, 183*, 674–693.

Cabrales, A., & Calvó-Armengol, A. (2008). Interdependent preferences and segregating equilibria. *Journal of Economic Theory, 139*, 99–113.

Camarinha-Matos, L. M., Afsarmanesh, H., Galeano, N., & Molina, A. (2009). Collaborative networked organizations—Concepts and practice in manufacturing enterprises. *Computers & Industrial Engineering, 57*, 46–60.

Cohen, R., Katzir, L., & Raz, D. (2006). An efficient approximation for the generalized assignment problem. *Information Processing Letters, 100*, 162–166.

De Farias, I. R., Johnson, E. L., Jr., & Nemhauser, G. L. (2000). A generalized assignment problem with special ordered sets, a polyhedral approach. *Mathematical Programming Ser. A, 89*, 187–203.

Ghosh, S., & Gagnon, R. J. (1989). A comprehensive literature-review and analysis of the design, balancing and scheduling of assembly systems. *International Journal of Production Research, 27*, 637–670.

Glover, F. W. (1994). Genetic algorithms and scatter search: Unsuspected potential. *Statistical Computing, 4*, 131–140.

Glover, F. W. (1996). Ejection chains, reference structures and alternating path methods for traveling salesman problems. *Discrete Applied Mathematics, 65*, 223–253.

Haddadi, S., & Ouzia, H. (2004). Effective algorithm and heuristic for the generalized assignment problem. *European Journal of Operational Research, 153*, 184–190.

Jeet, V., & Kutanoglu, E. (2007). Lagrangean relaxation guided problem space search heuristic for generalized assignment problems. *European Journal of Operational Research, 182*, 1039–1056.

Ko, H. S., & Nof, S. Y. (2012). Design and application of task administration protocols for collaborative production and service systems. *International Journal of Production Economics, 135*, 177–189.

Koçkesen, L., & Ok, E. A. (2000). Evolution of interdependent preferences in aggregative games. *Games and Economic Behavior, 31*, 303–310.

Kutanoglu, E., & Mahajan, M. (2009). An inventory sharing and allocation method for a multi-location service parts logistics network with time-based service levels. *European Journal of Operational Research, 194*, 728–742.

Laguna, M., Kelly, J. P., González-Velarde, J., & Glover, F. (1995). Tabu search for the multilevel generalized assignment problem. *European Journal of Operational Research, 82*, 176–189.

Lee, Y. H., Jung, J. W., & Jeon, Y. S. (2007). An effective lateral transshipment policy to improve service level in the supply chain. *International Journal of Production Economics, 106*, 115–126.

Lee, D., Stajkovic, A. D., & Cho, B. (2011). Interpersonal trust and emotion as antecedents of cooperation: Evidence from Korea. *Journal of Applied Social Psychology, 41*, 1603–1631.

Li, J., Moghaddam, M., & Nof, S.Y. (2016). Dynamic storage assignment with product affinity and ABC classification—a case study. *International Journal of Advanced Manufacturing Technology, 84*, 2179–2194.

Lorena, L. A. N., & Narciso, M. G. (1996). Relaxation heuristics for a generalized assignment problem. *European Journal of Operational Research, 91*, 600–610.

Lourenco, H. R. D., & Serra, D. (2002). Adaptive approach heuristics for the generalized assignment problem. *Mathware and Soft Computing, 9*, 209–234.

Moghaddam, M., & Nof, S. Y. (2014). Combined demand and capacity sharing with best matching decisions in enterprise collaboration. *International Journal of Production Economics, 148*, 93–109.

Moghaddam, M., & Nof, S. Y. (2015a). Balanceable assembly lines with dynamic tool sharing and best matching protocols—a collaborative assembly framework. *IIE Transactions, 47*, 1363–1378.

Moghaddam, M., & Nof, S. Y. (2015b). Real-time administration of tool sharing by best matching to enhance assembly lines balanceability and flexibility. *Mechatronics, 31*, 147–157. (Special Issue: "New Trends in Intelligent Manufacturing").

Moghaddam, M., & Nof, S. Y. (2015c). Best matching with interdependent preferences—Implications for capacitated cluster formation and evolution. *Decision Support Systems, 79*, 125–137.

Moghaddam, M., & Nof, S. Y. (2015d). Interdependent preferences—Definitions and impacts on team formation. In *The 2015 Industrial and Systems Engineering Research Conference*, Nashville, USA.

Moghaddam, M., & Nof, S. Y. (2015e). Collaborative location-allocation decisions in networked v-Organizations. Research memo, PRISM Center, Purdue University.

Moghaddam, M., & Nof, S. Y. (2016). Real-time optimization and control mechanisms for collaborative demand and capacity sharing. *International Journal of Production Economics, 171*, 495–506. (Special Issue: "Challenges for Sustainable Operations—Selected papers of ICPR 2013").

Morita, P. P., & Burns, C. M. (2014). Trust tokens in team development. *Team Performance Management, 20*, 39–64.

Nof, S. Y. (2007). Collaborative control theory for e-Work, e-Production, and e-Service. *Annual Reviews in Control, 31*, 281–292.

Owen, S. H., & Daskin, M. S. (1998). Strategic facility location: A review. *European Journal of Operational Research, 111*, 423–447.

Sahni, S., & Gonzalez, T. (1976). P-complete approximation problems. *Journal of the ACM, 23*, 555–565.

Seok, H., & Nof, S. Y. (2014). Collaborative capacity sharing among manufacturers on the same supply network horizontal layer for sustainable and balanced returns. *International Journal of Production Research, 52*, 1622–1643.

Sobel, J. (2005). Interdependent preferences and reciprocity. *Journal of Economic Literature, 43*, 392–436.

Tiacci, L., & Saetta, S. (2011). Reducing the mean supply delay of spare parts using lateral transshipments policies. *International Journal of Production Economics, 133*, 182–191.

Tomes, N. (1986). Income distribution, happiness, and satisfaction: A direct test of the interdependent preferences model. *Journal of Economic Psychology, 7*, 425–446.

Torabi, S. A., & Moghaddam, M. (2012). Multi-site integrated production-distribution planning with transshipment: a fuzzy goal programming approach. *International Journal of Production Research, 50*, 1726–1748.

Yagiura, M., Ibaraki, T., & Glover, F. (2004). An ejection chain approach for the generalized assignment problem. *INFORMS Journal on Computing, 16*, 133–151.

Yagiura, M., Yamaguchi, T., & Ibaraki, T., (1999). A variable depth search algorithm for the generalized assignment problem. In *Metaheuristics, advances and trends in local search paradigms for optimization* (pp. 459–471). Boston, MA: Kluwer Academic Publisher.

Yoon, S. Y., & Nof, S. Y. (2010). Demand and capacity sharing decisions and protocols in a collaborative network of enterprises. *Decision Support Systems, 49*, 442–450.

Yoon, S. Y., & Nof, S. Y. (2011). Affiliation/dissociation decision models in demand and capacity sharing collaborative network. *International Journal of Production Economics, 130*, 135–143.

Chapter 8
Frontiers in Best Matching

Abstract This chapter presents a brief summary of the emerging technologies influenced by best matching processes and decisions, along with major technical challenges of best matching models, algorithms, and protocols, and opportunities for future improvements. The purpose is to highlight the emerging thrusts in best matching theory, in terms of both theoretical developments and practical applications.

8.1 Emerging Technologies Dealing with Best Matching

Applications of best matching in practice are plentiful. The previous chapters of this book specified, formulated, and illustrated the remarkable impacts of best matching on the quality and performance of a wide variety of manufacturing, supply, logistics, communication, transportation, healthcare, and service systems. It was hypothesized, studied, and proven that a system with "mismatched" individuals is usually outperformed by an equivalent system with "best-matched" individuals. In other words, if every single element of a system is matched to its best peer(s) at the right time, e.g., suppliers, retailers, and customers in enterprise networks; jobs, machines, and sensors in factories; vehicles and routes in transportation systems; robots, humans, and orders in warehouses, the outcome will be a better system that addresses the emerging needs for higher flexibility, responsiveness, quality, scalability, fault tolerance, as well as timeliness and optimality of matching decisions.

With the growing trend towards distribution of operations and decisions—enabled by the emerging information, communication, and computing (cyber) technologies—the need for effective best matching frameworks, algorithms, and protocols is realized more than ever. The World Economic Forum recently published a list of top 10 emerging technologies of 2015, identified by a panel of 18 internationally renowned experts (Meyerson 2015). The list includes: (1) Fuel cell vehicles; (2) Next-generation robotics; (3) Recyclable thermoset plastics; (4) Precise genetic engineering techniques; (5) Additive manufacturing; (6) Emergent artificial intelligence; (7) Distributed manufacturing; (8) 'Sense and avoid' drones; (9) Neuromorphic technology; (10) Digital genome. Although some items in the

M. Moghaddam and S.Y. Nof, *Best Matching Theory & Applications*, Automation, Collaboration, & E-Services 3, DOI 10.1007/978-3-319-46070-3_8

World Economic Forum's list are out of the scope of this book, some can be significantly influenced by best matching—the cases that deal with *interactions* in distributed systems.

8.1.1 Connected Robots

The examples of intelligent warehouse management system and precision agriculture discussed in the previous chapters are two among several examples of the impact of best matching on the performance of emerging internetworked and autonomous robotics systems. Self-driving cars and autonomous drones are among other rising applications of next-generation robotics—by 2020, there will be 10 million self-driving cars on the road and the global unmanned aerial vehicle market will be worth more than 10 billion dollars (Business Insider Intelligence Estimates 2015).

Robots are becoming smaller and more collaborative, thanks to the evolution of cloud technologies that enable quick and efficient access to information and computing resources shared on the cloud. In this context, it is necessary for the autonomous robots to be able to solve different best matching problems in real-time and in an optimal fashion. Examples are order pickup and storage location decisions in warehouse systems; routing, clustering, and communication with sensors, robots, and base station in smart farms; crowd-sourcing, communicating with the road and traffic, and deciding with whom to interact, how, and when, in self-driving car applications.

8.1.2 Cloud Manufacturing

The notion of the Internet of Things along with the advances in manufacturing technologies (e.g., 3D printing) has substantially transformed the physical manufacturing world into cyber-physical and self-governing systems of embedded collaborative units. Emerging cloud technologies have enabled the decentralization and mass-customization of manufacturing processes, and transformation of supply networks of materials and finished goods into supply networks of information. This revolution has created a new manufacturing paradigm called *cloud manufacturing*, which is defined as

> a model for enabling ubiquitous, convenient, on-demand network access to a shared pool of configurable manufacturing resources (e.g., manufacturing software tools; manufacturing equipment; and manufacturing capabilities) that can be rapidly provisioned and released with minimal management effort or service provider interaction (Xu 2012).

This new paradigm and its treatment of *everything-as-a-service* have advanced a new cohort of best matching challenges; e.g., what services to move to the cloud, in what order, and when; how to structure and optimize interactions and collaborations among service providers and users (Wu et al. 2015). Chap. 7 presented detailed examples of such matching processes in collaborative manufacturing, supply, and service systems.

8.2 Technical Challenges of Best Matching

Several models, algorithms, and protocols were discussed in the previous chapters of this book, which are based on certain assumptions made to enhance their development, understanding, and application. This section summarizes some practical challenges associated with those best matching models, algorithms, and protocols, where the underlying assumptions of those matching models, algorithms, and protocols are not valid.

8.2.1 Efficient Computation and Communication

As discussed from the very beginning of this book, solving matching problems is functionally desirable but computationally expensive. Several algorithms were introduced and examined for solving various instances of best matching problems in efficient and optimal manners. In many applications, the runtimes of those optimization algorithms are less than the length of the period between two (required) consecutive runs; and thus, the algorithm can be executed in real-time, under the periodic review method (see Chap. 6), without significant obstacles. In applications where the gap between two consecutive executions is much smaller than the runtime of the algorithm, however, upholding the efficiency and optimality of decisions becomes a major challenge.

The real-time optimization with continuous review method (see Chap. 6) is the preferred approach in such situations; however, there is no guarantee of the computational efficiency of the current adaptation heuristics, especially if the ratio of the gap between two consecutive runs to the runtime of the algorithm is relatively very small. The other technical challenges that were disregarded in the previous chapters are *communication delay*, *failure*, and *cost*. All the matching models, algorithms, and protocols discussed so far are based on the assumption that communication is instantaneous, always available, and free of charge, which may not be true in practice. Therefore, an open area of research related to best matching is to improve the computation and communication models, concepts, and infrastructures to enhance the accuracy and timeliness of decisions in complex and rapidly evolving systems (e.g., a team of rescue robots; a large fulfillment center; a vehicle dispatching system).

8.2.2 Conflict and Error Detection and Prevention

A major assumption behind all the matching models, algorithms, and protocols discussed so far is that the matching process is free of any errors or conflicts. In theory, an error refers to any input, output, or intermediate result that does not meet the specifications, expectations, or objectives of a system. Examples related to best matching include incomplete or incorrect information (e.g., preference; capacity; requirement), malfunction of an individual in communication with its peers (e.g., a team of agricultural robots), or disobedience of an individual to follow its predefined protocol (e.g., a malicious or compromised software agent). A conflict, on the other hand, is the inconsistencies between the plans, activities, and objectives of a set of interacting individuals.

When two material handling robots take the same route at the same time, for example, they collide. The main difference between errors and conflicts is that an error may involve only one individual, while a conflict involves at least two individuals. As explained earlier in this book, any system that involves interactions involves best matching processes as well. According to the principle of Conflict and Error Detection and Prevention (CEDP) of the Collaborative Control Theory (CCT), the higher the rate of interactions in a system, the higher the probability of conflicts and errors (Chen and Nof 2007). Accordingly, the developed concepts, algorithms, and protocols of best matching must be tailored with respect to the CEDP principle of CCT, in order to minimize the rate of conflicts and errors in matching processes, prevent them from happening, and resolve them effectively.

8.2.3 Incentives for Collaboration

Distribution of decision-making authorities among individuals creates two different behaviors in their interactions with each other: Collaboration; Competition. The theoretical aspects of this issue were extensively discussed in Chaps. 1, 4, and 6 of this book. It was shown that when the local objectives of individuals are in line with each other, and with the global objectives of the system, they tend to collaborate; otherwise, they compete with each other. This extreme classification of behaviors simplifies the modeling and understanding of interactions in a distributed system; however, it may not be true in practice. Whether to collaborate or compete is not a binary decision—it depends on the level of individuals' *incentives* to collaborate or compete. Two individuals may collaborate and compete at the same time (see Fig. 4.1, Chap. 4), depending on the degree of correspondence between their local objectives. Therefore, the models, concepts, and protocols for distributed best matching must address the collaboration/competition incentives of individuals, and their impacts on the following (and similar) operational decisions (Polenske 2004; Prins 2010; Yoon and Nof 2011):

- What collaborative behaviors provide the best competitive advantages for an individual?
- What are the impacts of collaborative/competitive behaviors on the type and topology of the entire network?
- When should an individual join/leave/compete with a collaborative network?
- What are the benefits/costs of participating in a collaborative network?
- Why would an individual opt to remain in a collaborative network?
- What are the criteria and performance evaluation measures to characterize and assess the effectiveness and performance of a collaborative network?
- What are the asymmetries (e.g., power; resources) and mistrust issues (e.g., identity loss; mutual stereotyping)?
- How can collaborative relationships (e.g., trust; mutual learning) be developed?
- What are the strategies to deal with the uncertainties, openness, emergence, and complexity of a collaborative network?

8.2.4 Data Availability and Reliability

Throughout this book, all the modeling and experimental developments were based on the assumption that correct and clean data is always available. This is clearly not the case in many practical instances. For instance, an agent-based optimization mechanism was introduced in Chap. 7 for optimal tool sharing among collaborative workstations in real-time (see E_2, Sect. 7.3). The optimization mechanism is based on one primary assumption: The real-time performance of each workstation is known, instantaneously, at any given time. This assumption, although useful for abstracting and modeling the process, may not be valid in practical settings with considerable delays in data collection and communication together with errors in data.

In addition to such cases where data collection is time-consuming, there are some case in which accurate or complete data is even unavailable. For example, in applications such as image matching (image processing), sound matching, or genome (pattern) matching, the criterion (or criteria) by which matching preferences are formulated and the best match is identified may be very difficult or even impossible to quantify. Hence, the issues of data unavailability and incompleteness must be incorporated in the design of the best matching models, algorithms, and protocols discussed in the previous chapters of this book.

8.3 Summary

Future research will investigate the limitations of current best matching solutions, described in the previous section, considering key success factors for the next-generation manufacturing, supply, and service systems. Best matching models, algorithms, and protocols will enable real-time optimization and control of interactions in distributed systems, and enhance the intelligence of individuals for making timely and accurate decisions about *what to do*, *with whom to interact*, *how* and *when*. Realization of these goals requires (1) comprehensive characterization and taxonomy of distributed manufacturing and service systems, their requirements, constraints, and objectives; (2) assessment of the nature of interactions (e.g., competitive/cooperative), and development of intelligent and proactive mechanisms for handling controllable and uncontrollable behaviors; (3) design of effective decision-making networks through optimal distribution of decisional authorities (e.g., hierarchy/ heterarchy); (4) development of models and algorithms based on operations research, artificial intelligence, information technology, and cyber technology (i.e., analytics and informatics) that enable efficient and effective design, execution, and feedback.

Problems

8.1. Reflecting on this entire book: Can best matching be accomplished without collaboration? Explain why or why not, and illustrate your answer.

8.2. Consider a supply network of producers *P*, distributors *D*, retailers *R*, and consumers *C*.

(a) Identify three different types of mismatch between every pair, i.e., $\{P, D\}$; $\{D, R\}$; $\{R, C\}$; $\{C, \mathrm{P}\}$.
(b) Identify three different types of mismatch between all four sets.
(c) Match one best matching model M to resolving three different mismatch problems that you identified in (a, b).
(d) Based on this chapter, describe key emerging frontier issues in best matching that can influence your answer in (c).

8.3. Reconsider your answers in problem 8.2 for each of the following sets of the items supplied in the 4-tier supply network:

(a) Food and medical supplies.
(b) Billing invoices and receipts for interactions among the supply network parties.
(c) Signals in a 4-tier sensor network.
(d) Software apps (tools) needed right now for a current decision problem faced by a single consumer.
(e) Pre-stored advices to consumers who seek recommendation for a gift to purchase for a particular family member' life event.

8.4. In the answers to problem 8.2, how do mismatch problems influence:

(a) Responsiveness.
(b) Quality.
(c) Timeliness and optimality of matching decisions.

8.5. Which of the ten major emerging technologies identified in Sect. 8.1 rely on interactions that may cause mismatch, and which do not? Describe briefly some of the mismatch situations that may occur, and their related impacts.

8.6. Each of the five examples (a–e) in problem 8.3 is a particular case of the general best matching of {Problems} and {Solutions}. Describe how emerging developments presented in this chapter will influence this general best matching problem.

8.7. For each of the following cases: (i) What interactions require best matching? (ii) What problems of mismatch can be identified? (iii) What model of best matching from those presented in this book would be most appropriate?

(a) Connected robots cooperating on exploration mission.
(b) Connected vehicles sharing the same road (but otherwise having no related responsibilities).
(c) Connected vehicles deployed for shared distribution and delivery of orders to customers.
(d) Cloud manufacturing.
(e) A production cell with integrated 3D printers and inspection machines.

8.8. Describe several technical challenges of best matching, and illustrate how each of them influences one of the cases in Problem 8.7.

8.9. Compare and contrast the following three statements, illustrating your answer in the case of warehouse robots:

(a) Statement A: *Conflict and error detection and prevention methods applied in systems influence the matching results of best matching processes.*
(b) Statement B: *Conflict and error detection and prevention methods help the effectiveness of best matching in systems.*
(c) Statement C: *Conflict and error detection and prevention methods applied and integrated in the matching process itself improve the results of the matching process.*

8.10. Explain several ways to incent people and robots to accept the calculated, optimal or best matching recommendations related to them.

References

Business Insider Intelligence Estimates. (2015). 10 million self-driving cars will be on the road by 2020. URL: www.businessinsider.com/report-10-million-self-driving-cars-will-be-on-the-road-by-2020-2015-5-6. The drones report: Market forecasts, regulatory barriers, top vendors, and leading commercial applications. URL: www.businessinsider.com/uav-or-commercial-drone-market-forecast-2015-2

Chen, X. W., & Nof, S. Y. (2007). Error detection and prediction algorithms: Application in robotics. *Journal of Intelligent and Robotic Systems, 48*, 225–252.

Meyerson, B. (2015). Top 10 emerging technologies of 2015. *World Economic Forum*. URL: www.weforum.org/agenda/2015/03/top-10-emerging-technologies-of-2015-2

Polenske, K. R. (2004). Competition, collaboration and cooperation: An uneasy triangle in networks of firms and regions. *Regional Studies, 38*, 1029–1043.

Prins, S. (2010). From competition to collaboration: Critical challenges and dynamics in multiparty collaboration. *The Journal of Applied Behavioral Science, 46*, 281–312.

Wu, D., Terpenny, J., & Gentzsch, W. (2015). Economic benefit analysis of cloud-based design, engineering analysis, and manufacturing. *Journal of Manufacturing Science and Engineering, 137*, 040903–1–040903–9.

Xu, X. (2012). From cloud computing to cloud manufacturing. *Robotics and Computer-Integrated Manufacturing, 28*, 75–86.

Yoon, S. Y., & Nof, S. Y. (2011). Affiliation/dissociation decision models in demand and capacity sharing collaborative network. *International Journal of Production Economics, 130*, 135–143.

Erratum to: Best Matching Theory & Applications

Mohsen Moghaddam and Shimon Y. Nof

Erratum to:
M. Moghaddam and S.Y. Nof, *Best Matching Theory & Applications*, Automation, Collaboration, & E-Services 3, DOI 10.1007/978-3-319-46070-3

The original version of the book was inadvertently published without the following corrections:

In Chapter 4, the numbering and indent problems need to be fixed (similar to the other chapters).
In Chapter 6, in Page 133: “Step 4. (Revisited)” should be changed to read as “Step 2. (Revisited)”.
In Page 134: Table 6.2 should be placed in the middle of Step 3 (Page 133), right after “...collected during [0, 4)—as follows:” (“as follows” refers to that table)
In Chapter 7, Table 7.1: The entries of 4th column of 3rd row and 5th row have been aligned as per adjusting the width of the column.
“This book is dedicated to:” is missing in the dedication page and should be updated.
New series cover design to be updated.

The updated original online version for these chapters can be found at
http://dx.doi.org/10.1007/978-3-319-46070-3_4
http://dx.doi.org/10.1007/978-3-319-46070-3_6
http://dx.doi.org/10.1007/978-3-319-46070-3_7
http://dx.doi.org/10.1007/978-3-319-46070-3

M. Moghaddam and S.Y. Nof, *Best Matching Theory & Applications*, Automation, Collaboration, & E-Services 3, DOI 10.1007/978-3-319-46070-3_9

Index

A

Agent, 6, 9, 36, 64, 67, 144–146, 149, 150, 152, 154, 155, 163, 174, 180, 188
Allocation, 10, 11, 24, 28, 34, 65, 71, 72, 154, 158, 169, 174, 202, 203, 207, 210
Analogical reasoning, 19, 20, 38
Analogy, 8, 19, 20
Analytics, 125, 157, 226
Ant colony optimization, 110, 133, 136
Artificial Intelligence, 142, 154, 212
Assembly, 9, 10, 24, 30, 31, 47, 51, 52, 126, 159, 168, 182–184, 188, 192, 193
Assignment, 10, 14, 27, 28, 30, 48, 72, 157, 195
Association-dissociation, 12
Automated guided vehicle, 4
Automated storage/Retrieval system, 4, 68
Autonomy, 7, 67, 145, 154

B

Balanceability, 168, 182, 183, 187, 192
Best matching index, 11
Bottleneck, 27, 31, 57, 168, 183, 190, 193
Branch-and-bound, 83, 90, 98, 99, 118, 121

C

Cloud, 169, 222
Clustering, 10, 20, 169, 194, 196, 222
Collaboration, 2, 8, 9, 28, 30, 36, 65, 74, 75, 144, 168, 170–173, 176, 179, 180, 183, 187, 202, 210, 217, 224
Collaborative control theory, 12, 168, 196, 224
Collaborative networked organization, 29, 169
Communication, 2, 4, 11, 28, 36, 67, 130, 144, 154, 169, 195, 211, 221, 223, 224
Competition, 8, 9, 65, 74, 75, 110, 144, 182, 224
Complexity, 6, 11, 73, 77, 81, 90, 118, 142, 154, 158, 169, 196, 213
Computation, 130, 154, 155, 223
Conditions, 2, 6, 20, 22, 26, 29, 37, 41, 45, 49, 70, 102, 108, 125, 129, 174
Conflict, 2, 67, 73, 155, 174, 187, 203, 224
Continuous review, 135, 162, 196, 223
Control, 3, 8, 12, 29, 67, 70, 127, 138, 145, 147, 167, 174, 180, 193, 211, 226
Coopetition, 75, 213
Coordination, 12, 65
Cost of matching, 83
Criteria, 2, 3, 7, 23, 26–28, 34, 58, 59, 70, 105, 110, 118, 129, 179, 204
Cubic assignment problem, 56
Cyber, 6, 36, 221, 226

D

Defense, 8, 12
Deferred acceptance algorithm, 3, 87, 88, 118
Demand-capacity sharing, 167, 168, 172, 179, 180
Distributed decision-making, 63, 67, 72, 144
Distributed system, 6, 8, 11, 63, 67, 72, 76, 127, 147, 212, 226
Drone, 70, 71, 216, 222
Dynamic programming, 92, 95

E

Emergent lines of collaboration, 12, 30
Error, 158, 203, 224
Evolution, 6, 8, 43, 67, 118, 134, 169, 197
Exact algorithm, 83

M. Moghaddam and S.Y. Nof, *Best Matching Theory & Applications*, Automation, Collaboration, & E-Services 3, DOI 10.1007/978-3-319-46070-3

F
Factory, 2, 36
Formalism, 12, 21, 23, 27, 28, 32, 37, 64, 174, 212
Fuzzy optimization, 161

G
Genetic algorithm, 103, 121, 136, 142, 155
Goal programming, 28, 58
Greedy randomized adaptive search procedure, 103, 107
Group technology, 11

H
Healthcare, 8, 12, 36, 88, 169, 221
Heterarchy, 67, 73, 145, 188, 212
Heuristic, 12, 39, 82, 83, 102, 103, 118, 136–138, 169, 203, 207, 210
Hierarchy, 3, 67, 74
Hungarian method, 83, 86, 118

I
Incentive, 74, 212, 224
Informatics, 157, 160
Information and communication technologies, 6
Intelligence, 8, 68, 103, 127, 145, 154, 155, 211, 222
Interaction, 2, 4, 21, 36, 63, 67, 72, 74–77, 102, 145, 146, 154, 195, 212, 224, 226
Interdependent preferences, 34, 49, 53, 132, 169, 194
Internet of things, 222

L
Lagrangian relaxation, 83, 90, 91, 93, 105, 118
Layered matching, 49, 56
Local search, 103, 114
Location, 2, 4, 10, 11, 39, 54, 70, 72, 109, 126, 195, 204, 222
Logistics, 11, 157, 169, 201, 221

M
Machine, 2, 5, 8, 10, 20, 21, 23, 24, 31, 48, 64, 115, 116, 121, 147, 149, 152, 162, 194, 203
Manufacturing, 4, 8, 11, 19, 36, 52, 142, 154, 167, 211, 221–223
Material handling, 12, 65, 69, 70, 109, 126, 224
Mathematical model, 43, 49, 60, 167
Metaheuristic, 12, 81, 102, 103, 111, 206
Mismatch, 3, 226
Mixed-integer programming, 139
Multi-agent system, 64, 144, 146, 157, 168, 188

N
Natural selection, 9, 106, 156, 198
Nature, 1, 9, 21, 54, 63, 74, 75, 127, 145, 211
Negotiation, 9, 67, 87, 146, 153, 156, 170, 182, 212
Neighborhood search, 103, 107, 114, 203, 209

O
Optimization, 19, 83, 90, 98, 114, 125, 127, 130, 136, 147, 167, 187, 223
Order fulfillment, 179

P
Periodic review, 129, 135, 223
Population-based, 102, 111, 197
Power system, 12
Precedence relations, 2, 23, 37, 51, 61
Precision agriculture, 73, 78, 131, 222
Preferences, 2, 23, 24, 28, 34, 45, 52, 53, 55, 87, 118, 147, 155, 169, 194, 201
PRISM taxonomy, 11, 19, 21, 27, 30, 37, 40, 43, 59, 102, 160, 204
Prisoner's dilemma, The, 75
Process, 1, 2, 4, 6, 8, 9, 12, 21, 24, 28, 30, 31, 34, 49, 56, 69, 73, 87, 129, 132, 144, 149, 152, 156, 160, 168, 180, 183, 200, 224, 225
Procurement, 6, 36
Production, 3, 6, 9–11, 20, 23, 31, 70, 103, 169, 173
Progression, 12, 29, 30
Protocol, 3, 12, 20, 63, 64, 144, 146, 151, 171, 182, 212, 223

Q
Quadratic assignment problem, 53, 54

R
Radio frequency identification (RFID), 36
Real-time optimization, 29, 125, 128, 171, 211
Reliability, 26, 28, 144
Resource-constrained, 49, 50, 51, 204
Resource sharing, 51, 52, 118, 171, 173, 182

Robot, 2, 61, 65, 69, 71–73, 85, 121, 132, 211, 222, 224
Robustness, 38, 73
Robust optimization, 127, 128
Routing, 9, 20, 72, 222

S
Scheduling, 9, 11, 12, 14, 20, 30, 48, 126, 145
Sensor, 4, 8, 9, 29, 67, 131, 136, 139, 154, 188, 211, 222
Service, 3, 6, 20, 34, 37, 81, 144, 167, 170, 210, 221, 222, 226
Set partitioning, 118
Shop floor, 4, 12, 145, 147
Social network, 2, 201, 213
Stability, 3, 86, 168, 169, 173, 204, 210
Stochastic optimization, 128
Supply network, 8, 11, 23, 74, 159, 167, 168, 170, 171, 179, 222
System of systems, 64, 65

T
Tabu search, 103, 114, 118, 136, 169, 206, 208
Task administration protocol, 12, 125, 147, 174
Technology, 82, 212, 226
Telecommunication, 9, 64
Time, 3, 12, 20, 23, 29, 30, 32, 39, 49, 59, 71, 76, 83, 87, 107, 126, 130, 134, 138, 145, 147, 149, 152, 160, 168, 180, 183, 184, 186, 187, 190, 192, 207, 208, 224
Transportation, 1, 8, 12, 145, 157, 169, 211, 221

U
Utilization, 52, 168, 169, 182, 192, 204, 210

W
Warehouse management system, 68, 69, 84, 119, 222
Workflow, 45, 158, 159, 174

MIX
Papier aus verantwortungsvollen Quellen
Paper from responsible sources
FSC® C105338

If you have any concerns about our products,
you can contact us on
ProductSafety@springernature.com

In case Publisher is established outside the EU,
the EU authorized representative is:
Springer Nature Customer Service Center GmbH
Europaplatz 3, 69115 Heidelberg, Germany

Printed by Libri Plureos GmbH
in Hamburg, Germany